99-WWF-8

REDUCING PEAK RAINFALL-DERIVED INFILTRATION/INFLOW RATES—CASE STUDIES AND PROTOCOL

by:

M. Steven Merrill, Ph.D., P.E.
Andrew Lukas, M.S., P.E.
Brown and Caldwell

Cos Roberts, P.E.
RCT Engineering

Richard N. Palmer, Ph.D., P.E.
Nathan Van Rheenan
University of Washington

2003

Water Environment Research Foundation

The Water Environment Research Foundation, a not-for-profit organization, funds and manages water quality research for its subscribers through a diverse public-private partnership between municipal utilities, corporations, academia, industry, and the federal government. WERF subscribers include municipal and regional water and wastewater utilities, industrial corporations, environmental engineering firms, and others that share a commitment to cost-effective water quality solutions. WERF is dedicated to advancing science and technology addressing water quality issues as they impact water resources, the atmosphere, the lands, and quality of life.

For more information, contact:
Water Environment Research Foundation
635 Slaters Lane, Suite 300
Alexandria, VA 22314-1177
Tel: (703) 684-2470
Fax: (703) 299-0742
www.werf.org
werf@werf.org

This report was co-published by the following organizations. For non-subscriber sales information, contact:

Water Environment Federation
601 Wythe Street
Alexandria, VA 22314-1994
Tel: (800) 666-0206
Tel: (703) 684-2452
Fax: (703) 684-2492
www.wef.org
pubs@wef.org

IWA Publishing
Alliance House, 12 Caxton Street
London SW1H 0QS, United Kingdom
Tel: +44 (0) 20 7654 5500
Fax: +44 (0) 20 7654 5555
www.iwapublishing.com
publications@iwap.co.uk

Library of Congress Catalog Card Number: 2003106628

WERF ISBN 1-893664-76-7 IWAP ISBN: 1-84339-652-1 WEF ISBN: 1-57278-213-7

Brown and Caldwell

ACKNOWLEDGMENTS

Report Preparation

Principal Investigator:
M. Steven Merrill, Ph.D., P.E.
Brown and Caldwell

Project Team:
Matthew Davis, M.S., P.E.
Anthony M. Dubin, M.S.E., P.E.
Lisa Johnston
Andrew Lukas, M.S., P.E.
Heidi McEllistrem, P.E.
Julie McMullin, M.S., E.I.T.
Brown and Caldwell

Cos Roberts, P.E.
RCT Engineering

Richard N. Palmer, Ph.D., P.E.
Nathan Van Rheenan
University of Washington

Project Subcommittee

Tyler Richards
Gwinett County, Georgia

Peter Keefe, P.E.
ADS Environmental Services

Dan Murray
United States Environmental Protection Agency

Robert Swarner, P.E.
King County Department of Natural Resources

Water Environment Research Foundation Staff

Research Program Director: Jeff Moeller, P.E.
Project Manager: Roy Ramani, Ph.D.

The project team would like to thank all of the communities with projects that were considered for case studies. These communities are:

- City and County of Honolulu, Hawaii
- City of Issaquah, Washington
- City of Kent, Washington
- City of Milwaukee, Wisconsin
- City of Oak Creek, Wisconsin, and the Oak Creek Water and Sewer Utility
- City of Olympia, Washington
- Clean Water Services, Hillsboro, Oregon
- King County Department of Natural Resources, King County, Washington
- Lacey, Olympia, Tumwater, and Thurston (LOTT) Wastewater Partnership
- Lower Paxton Township, Harrisburg, Pennsylvania
- McCandless Township Sanitary Authority, Pittsburgh, Pennsylvania
- Milwaukee Metropolitan Sewerage District (MMSD), Milwaukee, Wisconsin
- Susquehanna Township, Harrisburg, Pennsylvania

In addition, the project team would like to thank Paula Aresenault, Earth Tech, Fukunaga & Associates, Inc., Sterns & Wheler Companies, and Gibbs and Olson, Inc. for their data contributions.

ABSTRACT AND BENEFITS

Abstract:

The purpose of this research project was to examine construction projects that successfully reduced rainfall-derived infiltration and inflow (RDII). The research began with a literature review of all published records describing RDII removal projects. Initially, the review identified many projects; however, subsequent evaluation of available information revealed several important realities.

First, most RDII removal projects in the country go undocumented. Second, of the RDII removal project summaries that have been published, few provide good data. Third, data gathering and analyses for this report were hampered by lack of documentation, lost or unavailable monitoring data, and weaknesses in monitoring techniques.

Projects with sufficient information describing the conditions before the RDII removal project received a detailed analysis. In all, the case studies from the six agencies presented in this report 1) document before-and-after RDII levels, 2) quantify the RDII reduction achieved, and 3) describe the cost-effectiveness of the removal. As more than one project was documented for some agencies, a total of 12 RDII projects received review. The projects examined include both those that "successfully" reduced RDII and those that did not. In general, the conclusion was that utilities run the risk of not removing significant RDII unless they address private sewer laterals. The only cases examined where significant RDII was removed were those that addressed private sewers or inflow sources.

A major recommendation of this report was to establish a standard reporting protocol for documenting RDII removal projects. This protocol details what specific information should be provided when documenting an RDII removal project and how to perform an evaluation of RDII removal effectiveness. The protocol touches on issues ranging from flow meter maintenance to statistical analysis of flow data. By using the proper documentation recommended in this report, communities allow others who are preparing to perform similar work to examine and utilize the results of any RDII removal project.

Benefits:

- Describes in detail the approaches and results for a variety of RDII removal projects
- Provides an analytical approach for determining the reduction in peak and average flow caused by sewer rehabilitation for RDII removal
- Includes a literature search of RDII removal projects and technologies
- Provides an effectiveness evaluation reporting protocol for RDII removal projects
- Recommends a nationally coordinated effort for storing and distributing the results of RDII removal projects that are summarized according to the protocol

Keywords: infiltration, inflow, sewer rehabilitation, sanitary sewer overflows, flow monitoring, modeling, statistics

TABLE OF CONTENTS

LIST OF TABLES

LIST OF FIGURES

LIST OF ACRONYMS

ADWF	Average Dry Weather Flow
API	Antecedent Precipitation Index
CCTV	Closed-Circuit Television
CFS	Cubic Feet per Second
CIPP	Cured-in-Place Pipe
CMUD	Charlotte-Mecklenburg Utility Department
DIP	Ductile Iron Pipe
ENR	Engineering News-Record
gpad	Gallons per Acre per Day
gpd	Gallons per Day
gpm	Gallons per Minute
GWI	Groundwater Infiltration
HDPE	High Density Polyethylene
I/I	Infiltration and Inflow
idm	Inch-Diameter-Mile
LF	Lineal Feet
LOTT	Lacey, Olympia, Tumwater, and Thurston
LP3	Log Pearson Type III
LSSES	Limited Sanitary Sewer Evaluation Survey
mgd	Million Gallons per Day
MMSD	Milwaukee Metropolitan Sewerage District
MOP	Manual of Practice
MWRDGC	Metropolitan Water Reclamation District, Greater Chicago
NCDC	National Climatic Data Center
NOAA	National Oceanic and Atmospheric Administration
O&M	Operation and Maintenance
PROW	Public Right-of-Way
PVC	Polyvinyl Chloride
RDII	Rainfall-Derived Infiltration and Inflow
RFP	Request for Proposal
SFMA	South Fayette Township Municipal Authority
SPTSD	South Palos Township Sanitary District
SSES	Sanitary Sewer Evaluation Survey
SSO	Sanitary Sewer Overflow
SWMM	Stormwater Management Model
USA	Unified Sewerage Agency
USDA	United States Department of Agriculture
U.S. EPA	United States Environmental Protection Agency
USGS	United States Geological Survey
UW	University of Washington
VCP	Vitrified Clay Pipe
WLSSD	Western Lake Superior Sanitary District

WW	Wet Weather
WWF	Wet Weather Flow

EXECUTIVE SUMMARY

Evaluation of rainfall-derived infiltration and inflow (RDII) removal in separated sewer systems is becoming increasingly important. Collection and treatment systems are exceeding design capacities well ahead of their original design life, resulting in surcharging, backups, bypasses, and reduced treatment efficiency. Discharge regulations are becoming more stringent, requiring reductions in sewer overflow events and more extensive treatment (e.g., larger facilities). Operation and maintenance costs continue to climb as greater-than-designed flows result in excessive wear, higher power demands, and operation outside the optimum range of efficiency.

Although the need for RDII reduction is increasing, there is uncertainty as to what approach will achieve significant improvement. The literature is replete with case studies where RDII removal is described as a success; however, conventional wisdom says little can be accomplished.

This report is designed to assist those confronted with the need to reduce peak RDII flows:

- Chapter 1.0 discusses the purpose of this report, which is to summarize data from various RDII removal projects and compare the RDII reduction rates achieved for each.
- Chapter 2.0 reviews the literature covering the analytical approaches for RDII evaluation and projects that document RDII removal.
- Chapter 3.0 discusses the analytical methods used for quantifying RDII and recommends the most appropriate approach.
- Chapter 4.0 discusses the approaches used for determining the reduction in peak and average RDII flows following sewer rehabilitation, and recommends the most appropriate approach.
- Chapter 5.0 describes the approaches used to reduce peak-hour RDII in various RDII removal projects. A variety of projects was examined—both those that "successfully" reduced peak RDII flow rates and those that did not. In a consistent fashion, the RDII reduction results achieved in each were examined and documented.
- Chapter 6.0 provides an effectiveness evaluation reporting protocol for RDII removal projects. The authors believe that if communities followed a standard reporting protocol, they would enable others to benefit from their RDII removal project reports.

Obtaining case study data for this report proved difficult. The reporting protocol provided here, therefore, defines the procedures for furnishing adequate information on completed RDII removal projects. Topics include defining basin characteristics, providing details of the RDII reduction construction project, and performing post-construction analysis of flow-monitoring data to determine RDII reduction effectiveness. A standard reporting protocol would not only completely describe the project and but also yield a valuable resource to communities researching the best method for evaluating/reducing RDII. Previously performed projects could be evaluated and used to determine what would work for their particular community.

ES.1 Conclusions

This report examined in detail the peak RDII flows in 12 projects conducted by six utilities. The rehabilitation methods ranged from manhole rehabilitation alone to complete replacement of all pipes in the study basins, including the sewers in the public right-of-way (PROW) as well as private sewer laterals. The results of the examinations can be characterized as follows:

1. Rehabilitation of just the sewers in the PROW may provide little reduction in peak hour RDII flows. One study showed a 17% reduction in peak-hour RDII flows when replacements included the portion of the building laterals in the PROW. The other projects of this type that were examined resulted in a 5% or less reduction in peak RDII flows.
2. As a corollary to #1, projects that addressed the private sewers from the PROW to the house achieved significant reduction in peak-hour RDII flows. Reductions of 50–70% were documented.
3. Rehabilitation of public sewers may have benefit in reducing the overall volume of RDII. Reductions in peak 24-hour average RDII volumes ranged from 2–30%. Reductions in peak monthly average flows ranged from 2–65%. Reduction in the total volume of RDII but not the peak indicates that the base of the hydrographs has been shortened. This suggests that infiltration from the groundwater entering public sewers can be reduced significantly in certain conditions (depending on the overall groundwater conditions). Although reductions in peak-day and peak-month RDII volumes benefit wastewater treatment facilities, they do not necessarily benefit the conveyance system.
4. The exception to the general finding (noted in #1) that rehabilitation activities in the PROW provide little or no peak RDII reduction was a manhole rehabilitation project in Milwaukee, Wisconsin. This project apparently achieved a 45% reduction in peak RDII flows through manhole rehabilitation alone. The circumstances suggest that attention to manholes as inflow sources in conditions where ground conditions reduce the impact of groundwater may achieve significant results. Manhole rehabilitation in other case studies, however, did not achieve the same results.

This report compares the case study projects in terms of pre-construction RDII, residual RDII, and cost-effectiveness of RDII removal, expressed in dollars per gallons per day. Chapter 5.0 discusses construction costs and effectiveness in terms of dollars per reduction in peak RDII rate. The finding was that cost-effectiveness varies as pre-rehabilitation RDII level increases (see Figure ES-1).

With enough projects representing a spectrum of RDII removal projects, it would be possible to generally determine what is "cost-effective" RDII removal. As presented in this report, however, insufficient data exists in the public domain for making such determinations. If more projects are documented in accordance with a protocol consistent with this report, it will be possible at some future time to better identify what is "cost-effective" RDII control. The report, however, does identify approaches to removal that have the greatest opportunity for success.

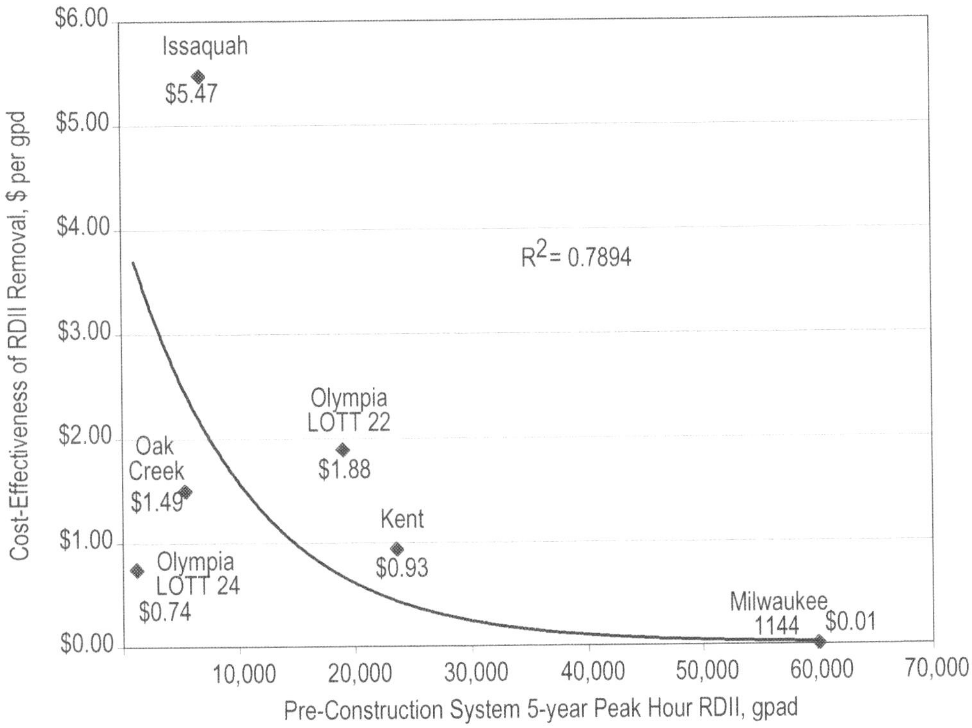

Figure ES-1. Cost-Effectiveness Comparison Curve (gpad) for Select RDII Case Study Projects.

ES.2 Recommendations

This study has identified approaches that have the highest probability for successfully reducing peak RDII flows. The limited sampling of projects prevents the making of generalized statements on cost-effectiveness. The recommendation is to accumulate additional data, particularly on large basins, over time. To facilitate drawing accurate conclusions from new data, it is further recommended that the reporting protocol described in Chapter 6.0 be followed. To further facilitate review of this data by third parties, it is recommended that agencies archive data in a manner to recover it easily.

Currently, several utilities are undertaking significant pilot projects that could greatly expand the knowledge base for RDII reduction. No national organization, however, is attempting to provide for a non-biased collection of infiltration and inflow reduction project data. It is further recommended, therefore, that an effort be undertaken to establish an RDII reduction project tracking system following the recommended protocol. Such an effort may be a combined project of WERF, the United States Environmental Protection Agency, the American Society of Civil Engineers, and the American Public Works Association.

Table ES-1 presents the projects, elements included, and results in terms of estimated reduction in the peak RDII flows having a five-year recurrence interval.

Table ES-1. Summary of Rehabilitation Techniques and RDII Removal for Each Project.

Project	*Type of Project*					*5-Year Peak-Hour RDII (gpad)*		
	Manhole Rehabilitation	*House Lateral Rehabilitation*	*Sewer Main Rehabilitation*	*Roof Drain Disconnection*	*Foundation Drain Disconnection*	*Pre-Construction*	*Post-Construction*	*Percent Reduction*
Oak Creek					√	5,474	2,851	48%
Honolulu—Basin A	√		√			8,063	7,666	5%
Honolulu—Basin B	√		√			12,468	12468	nil
Lacey, Olympia, Tumwater, and Thurston (LOTT)— OL22 Phase I		√ (lower)	√			19,047	15,807	17%
LOTT—OL22 Phase II		√ (upper)	√	—[a]	—[a]	15,807	7,421	53%
LOTT—OL24		√				1,219	573	53%
King County—Kent	√	√	√			23,680	5,300	78%
King County—Issaquah	√	√	√			6,877	3,390	51%
McCandless—Moreland Road Basin		√		√		—	—	nil
McCandless—Woodland Road Basin	√	√		√		—	—	nil
McCandless—Willoughby Road Basin		√	v			—	—	nil
Milwaukee Basin 1144	√					60,190	27,243	45%

CHAPTER 1.0

Report Purpose

To comply with emerging sanitary sewer over flow (SSO) regulations expected from the United States Environmental Protection Agency (U.S. EPA) in the near future, many municipal agencies want to know the most effective ways to remove rainfall-derived infiltration and inflow (RDII) within their systems. The questions for these agencies are "How much must be removed?" and "What is the best way to do it?" For many communities, the decisions on what sort of RDII removal project to undertake can be daunting, The most valuable information for these communities is what they can learn from other communities with successfully completed RDII removal projects.

The purpose of this report is to summarize date from various RDII removal projects and compare the RDII reduction rates achieved for each. The report includes an analysis of the various types of RDII removal projects completed in several cities and an evaluation of the effectiveness of the RDII removal for each. The project summaries include the project background (i.e., type of the system, amount of sewer system and characteristics, size of basin) and project details (i.e., construction/rehabilitation method used, portion of basin repaired, project costs).

Instead of simply repeating the reported RDII reduction rates achieved by each project analyzed, this research effort included the establishing of an accepted means of evaluating before-and-after flow and rainfall date for each project. In so doing, the effectiveness of the rehabilitation in reducing peak RDII flow rates in the system could be objectify compared among the projects.

In addition to documenting the RDII removal projects in consistent fashion, this report provides a standard reporting protocol for documenting future projects. Although the results of the analysis in this report provide some insight into what defines cost-effective RDII reduction, the need for more data clearly exists.

CHAPTER 2.0

Literature Review and Identification of Candidate RDII Removal Projects for Study

2.1 Literature Review

Sanitary sewers are designed to carry waterborne wastes from municipal, industrial, public, or private locations containing very little inadvertent stormwater, surface water, and groundwater (*Existing Sewer Evaluation and Rehabilitation—WEF Manual of Practice (MOP)*, 1994). Many sanitary sewers, however, experience significant increases in flows during and after storm events. When flow in sanitary sewers exceeds the capacity of the conveyance system, untreated or partially untreated sewage may be discharged from the system as either overflow or backflow. SSOs carry untreated pollutants into watercourses; backflows affect the homes of collection system customers by causing basement flooding (Brock, 1996). These occurrences, which most often are associated with wet weather (WW) events, can also occur during dry periods (Golden, 1996). In cities like Dallas, even small storms can cause SSOs (Sieger and King, 1995).

The primary causes of SSOs are RDIIs. Inflow is stormwater that enters a sanitary sewer system through catch basins, roof leaders, cleanouts, foundation drains, sump pumps, open manholes, and cellar, yard, and area drains. Infiltration is water that enters a sewer system from the ground through defective pipes, pipe joints, damaged lateral connections, and manhole walls (*Existing Sewer Evaluation and Rehabilitation—WEF MOP*, 1994; Sullivan et al. 1977). Damaged lateral connections are thought to be a major contributing factor to infiltration in sanitary sewers (Swarner and Thompson, 1994).

An early response to the problem of RDII was to design for peak storm flows and their accompanying inflows. In the 1950s, as much as eight times the normal dry weather peak flow capacity was provided. After a few years, however, sewers typically ran full during storm periods due, in part, to RDII and increased population and water consumption (Brock, 1996; Field and O'Connor, 1997). RDII from local agency sewer systems is driving the need for major conveyance and treatment expansion over the next 30 years (Swarner and Klusman, 1995).

Because of the size of many sanitary sewer systems, it is often far less costly to use alternatives to sewer rehabilitation for RDII and SSO control than to replace large sections of sewer line. In addition to the high costs associated with sewer rehabilitation, the time required to complete projects, and the general disruption caused, the reduction of RDII following rehabilitation is not always predictable or considered cost effective (Field and O'Connor, 1997).

Swarner and Klusman predict that although an effective RDII reduction program in King County, Washington, could reduce capital costs by more than $200 million, the overall reduction in peak RDII could be as little as 15–30% (Swarner and Klusman, 1995).

In discussing a control strategy for SSO, Field and O'Connor (1997) noted that lateral connections could contribute up to 80% of the infiltration load. Lateral connection rehabilitation, however, may be impractical and economically infeasible because of the number of connections. Because of the seemingly unpredictable nature of RDII reduction following sanitary sewer rehabilitation, it is increasingly difficult for wastewater utilities to confidently implement effective and economical RDII reduction programs.

2.1.1 Sanitary Sewer System Rehabilitation Management

To effectively manage a sanitary sewer system, it is essential to inventory the system thoroughly and to ascertain the condition and performance of all components—from individual pipelines to the complete collection system program approach (Stevens, 1994). The rehabilitation strategy for a sanitary sewer system, at any level, involves this same kind of investigation, whereby both the structural condition and hydraulic performance of the entire system is studied. This can lead to a long-term planning approach that encompasses the overall current and future needs of the entire system *(Existing Sewer Evaluation and Rehabilitation—WEF MOP*, 1994). Many cities are now implementing sewer construction and rehabilitation projects designed both to improve current conditions and ensure upgrades to meet future demands (Forbes, 1993).

2.1.2 Components of Flow Measurement in System Rehabilitation Management

In sewer system evaluation, flow measurement is undertaken to define the variation of flow components with time or to define peak or minimum flow conditions. Typically, a temporary flow-monitoring program is installed to measure flows at selected locations. Because flow measurements may be used in the design of capital improvements, it is critical to make accurate measurements. Sanitary sewer system flow comprises base flow, direct inflow, and infiltration. The primary objective of flow monitoring is to identify and quantify these three components *(Existing Sewer Evaluation and Rehabilitation—WEF MOP,* 1994):

1. ***Base Flow.*** With varying degrees of accuracy, base flow can be determined in several ways. Wastewater data—adjusted for seasonal peaks, irrigation, and unmetered connections—is often used for calculating base flow. Other methodologies include subtracting minimum flow rates (measured to estimated infiltration rates) from metered dry weather flow (*Existing Sewer Evaluation and Rehabilitation—WEF MOP*, 1994) and evaluating residential, commercial, and industrial water consumption upstream of the metering location (Swarner and Thompson, 1994).
2. ***Direct Inflow.*** In small basins, direct inflow may express itself as immediate response to rainfall. Since even small storms contribute to direct inflow, it can be measured over several small storms (Swarner and Thompson, 1994). As the basin size increases, however, it becomes increasingly difficult to separate direct inflow from other rapidly

responding sources such as foundation drains. If direct inflow is expected even in small basins, it is prudent to verify its existence through smoke testing and dye testing. Two peak rates of direct inflow should be established. An instantaneous peak direct inflow rate is required to size pumping stations, interceptors, and other equipment that must handle peak WW flows (WWFs) (Conklin, 1981). A peak daily flow rate (24-hour average) may be used to size flow storage and settling devices (*Existing Sewer Evaluation and Rehabilitation—WEF MOP,* 1994). Attempts have been made to correlate peak inflow rates with rainfall intensity with some success (Swarner and Klusman, 1995). If, however, peak inflow rates will be extrapolated from gauging data, care must be taken to ensure that proper equipment is in use if the monitoring site surcharges or overflows.

3. ***Infiltration.*** Measuring infiltration as a component of flow is a much more complex process than direct inflow measurement. For accurate calculations, one requires data from a number of sizable storms, spatial rainfall data, and flow data throughout the system (Swarner and Thompson, 1994). Infiltration volumes must be adjusted for seasonal fluctuations of groundwater levels. It is necessary, therefore, to measure both dry weather and peak WW infiltration. Dry weather infiltration is calculated by subtracting base flow from the flow measured at a particular site after several weeks of dry weather (*Existing Sewer Evaluation and Rehabilitation—WEF MOP*, 1994). Ideally, rates of WW infiltration or rainfall-dependent infiltration should be determined by correlating groundwater elevation data to peak WWFs in sanitary sewer systems during periods of high groundwater conditions (Swarner and Thompson, 1994).

Other sewer system and RDII evaluation methods that may be included with flow monitoring are:

- ***Precipitation Measurement.*** Correlates rainfall with flow-monitoring data to determine the amount of inflow entering the sanitary sewer system. Information provided will include rainfall intensity, total volume per event, and duration of the event *(Existing Sewer Evaluation and Rehabilitation—WEF MOP*, 1994).
- ***Groundwater Gauging.*** Locates the level of groundwater in soils and identifies variation *(Existing Sewer Evaluation and Rehabilitation—WEF MOP*, 1994).
- ***Smoke Testing.*** Detects RDII sources in a relatively quick and inexpensive way. Negative findings, however, do not necessarily prove that RDII does not exist *(Existing Sewer Evaluation and Rehabilitation—WEF MOP*, 1994; Wilmut and Sanders, 1990).
- ***Dye Testing.*** Identifies specific defects that can contribute RDII during rainfall or snowmelt. This rainfall simulation technique can also be effective in quantifying the amount of RDII entering a section of sewer under a controlled runoff situation *(Existing Sewer Evaluation and Rehabilitation—WEF MOP*, 1994; Wilmut and Sanders, 1990).

Television inspection—a means of visually identifying areas having excessive RDII—can be particularly useful in rapid verification of RDII variation (Wilmut and Sanders, 1990). Obviously, the inspections have to take place during WW to supply these observations.

The consensus among North American engineers and regulators is that RDII reduction following sanitary sewer rehabilitation is largely unsuccessful (Keefe, 1998) and does not justify the expense. This, however, may not be the case. The lack of success can be attributed to inaccurate flow measurement, surcharging, and complexity of flow balances (Conklin, 1981; Wilmut and Sanders, 1990). Even so, it is agreed that flow monitoring is a valuable and valid tool for RDII investigation and, despite its limitations, is essential to proper sanitary sewer management practice. In fact, many public utilities have begun installing permanent flow meters as standard practice, allowing city engineers to make predictive decisions, rather than operate in a reactive fashion (*Worldwater Env. Eng.*, 1996).

2.1.3 RDII Reduction Following Sanitary Sewer Rehabilitation

Keefe argues that the perception that sewer rehabilitation does not significantly reduce RDII is a result of inadequate reporting of successful projects. Most engineers, he notes, believe that sewer rehabilitation can only remove up to 35% of RDII. In a survey of 91 sewer rehabilitation projects, however, Keefe noted an average reported peak-rate RDII reduction of 49% (1998).

Representative examples of sanitary sewer rehabilitation projects that have measured reductions in RDII are noted and summarized below. For the most part, the reduction estimates reported in these studies are based on reduction in the percentage of total RDII volume during a rainfall event divided by the total rainfall volume in that event. Such reductions are often significantly larger than the corresponding reduction in peak RDII flows.

2.1.3.1 Metropolitan Waste Reclamation District, Greater Chicago (MWRDGC)

The MWRDGC owns intercepting sewers and water reclamation plants; however, 125 local communities own the local sewers that collect the sewage. The agency, having been beset by pollution problems caused by excessive RDII in the local sanitary sewer systems, required the local communities to conduct their own sewer evaluation and rehabilitation programs to remove excess RDII in compliance with U.S. EPA requirements (Macaitis and Kuhl, 1994; Bergman and Hardeman, 1991). Rehabilitation projects and incidental RDII reductions for several representative communities within the MWRDGC are summarized in the next four paragraphs.

2.1.3.2 Des Plaines, Elk Grove Village, Winnetka, and Hoffman Estates (1987)

Public sector rehabilitation work included manhole repairs, cross-connection removal, sewer line replacements, sewer line grouting, and sewer lining. Private sector rehabilitation included downspout and sump pump disconnection. Reported RDII reduction ranged from 32–68%, which was very close to predicted inflow reductions. The overall effectiveness level was rated between 97% and 114% (Lambert and Hollenbeck, 1987).

2.1.3.3 Melrose Park (1990)

Public sector work included use of fold-and-form trenchless pipe lining to repair 29,000′ of 8″ polyvinyl chloride (PVC). RDII was reportedly reduced by 49% (*Water Eng. Management*, 1990).

2.1.3.4 Village of Wilmette (1992)

Public sector rehabilitation work included manhole repairs, sewer line replacements, and sewer line grouting. Private sector rehabilitation included downspout and sump pump disconnection. Rehabilitation greatly reduced peak WWF; however, basement flooding was not eliminated. It was believed that indirect cross-connections between private storm sewer laterals and sanitary lateral were the primary source of high WWFs. Replacing portions of the sanitary laterals, however, did not appreciably reduce flows in test areas (Rowlett and Kelgard, 1992).

2.1.3.5 Northbrook (1983)

Manhole rehabilitation on a fraction of the town's manholes reportedly reduced total system RDII by up to 10% (Driver, 1983).

2.1.3.6 Metropolitan Department of Water and Sewerage Services, Nashville, Tennessee (1994–1997)

Public sector rehabilitation work included manhole repairs, cross-connection removal, sewer line replacements, sewer line grouting, and sewer lining, as well as an overall expansion of collection and treatment facilities (Dillard et al. 1993). RDII reduction in more than 12 measured basins averaged 44% (Keefe, 1998). RDII removal was reported as a reduction in percentage of rainfall captured (total volume) and not peak RDII flow rate.

2.1.3.7 Charlotte-Mecklenburg Utility Department (CMUD), North Carolina (1996)

Public sector rehabilitation work included manhole repairs, cross-connection removal, sewer line replacements, sewer line grouting, and sewer lining, as well as increasing storage and gravity sewer and pumping capacity (Miles et al. 1996). RDII reduction in four measured basins averaged 29 percent (Keefe, 1998).

2.1.3.8 Cordova, Alaska (1993)

Public sector rehabilitation work included manhole repairs, sewer line grouting, sewer lining, and point repairs. RDII reduction over eight measured basins averaged 70% (Stephl and Maciariello, 1993).

2.1.3.9 Bryan, Ohio (1996)

Public sector rehabilitation work comprised manhole rehabilitation only. RDII reduction averaged 50% (Rosendaul et al. 1996).

2.1.3.10 Unified Sewerage Agency (USA) of Washington County, Oregon

As part of a consent decree, USA was mandated to stop system overflows. Since USA started its RDII reduction program in 1990, an optimal cost removal rate of less than $2.00 per gallon became standard. The basins are evaluated using flow meters, and the amount of infiltration and storm-induced RDII for each basin is determined. If the expected reduction is greater than 70%, USA will estimate the cost of rehabilitating areas of the basin. If the cost of rehabilitation to achieve predicted RDII removal is less than $2.00 per gallon of RDII removed, the project will go forward. If not, USA will look at the next project on the list. Since 1990, USA has monitored the effects of sanitary sewer rehabilitation on infiltration and inflow (I/I) reduction.

Rehabilitation projects and RDII reductions for four communities within USA are summarized in the next four paragraphs. For each community, analysis methods used included flow components, precipitation measurement, smoke testing, visual inspection, and cameras. The modeling method used was the percentage of rainfall volume (R-factor) to quantify RDII in the system. Rehabilitation methods used included relining/sliplining, service lateral rehabilitation, inflow control, manhole rehabilitation, catch basins, and roof drain disconnection.

2.1.3.11 Banks (1991–1999)

Rehabilitation work included sliplining and side sewer repairs. Estimated RDII reduction was approximately 60%.

2.1.3.12 Cedar Hills (1991)

Rehabilitation work included sliplining and manhole and side sewer repairs. Estimated RDII reduction was approximately 74%.

2.1.3.13 Cornelius (1991–1999)

Rehabilitation work included sliplining and side sewer repairs. Estimated RDII reduction was approximately 50%.

2.1.3.14 Forest Grove (1991–1992)

Rehabilitation work included sliplining and manhole and side sewer repairs. Estimated RDII reduction was approximately 86%.

2.1.3.15 South Palos Township Sanitary District (SPTSD), Cook County, Illinois

The SPTSD, a small residential community, has experienced major RDII since its sewer system was originally constructed in the early 1970s. Although the SPTSD contains only 11 miles of sanitary sewer mains serving a mere 562 homes, two churches, and a school, the system regularly experiences peak flows in excess of 1.70 million gallons per day (mgd) for 1″ rainfalls. A comprehensive sewer evaluation was performed to determine dry weather flows and WWFs. Inspections were performed to identify sewer mains and laterals that were subject to infiltration. The reduction in peak RDII flow for a half-inch rainfall was calculated by evaluating flows before and after the sewer improvements. A 60% reduction in daily infiltration was achieved by cured-in-place pipe (CIPP) lining of approximately 25,000 lineal feet (LF) of sewer main (43% of system) and 150 sewer service laterals (27% of total) (Fundich and Kiest, 1999).

2.1.3.16 Rock River Water Reclamation District, Rockford, Illinois

The Rock River Water Reclamation District (the District) provides wastewater collection and treatment services to more than 200,000 residents in Rockford, Illinois, and the surrounding communities. The District implemented a phased multiyear program to mitigate basement backups after intense storm events in the Harrison Alpine service area. The program included identification and rehabilitation of public sector inflow sources, which accounted for only about 25% of the total system inflow, as well as identification and removal of 118 private sector inflow sources. Post-rehabilitation flow monitoring confirmed that 10-year storm protection from overflows and basement backups was achieved. Just as important, the project will serve as a model for implementing a private property RDII source identification and elimination program throughout the District (Hollenbeck and Rieger, 2001).

2.1.3.17 Paxton Creek Interceptor, Harrisburg, Pennsylvania

The Paxton Creek Interceptor—serving Susquehanna Township and Lower Paxton Township and located in Harrisburg, Pennsylvania—experiences SSOs during extreme WW conditions. To comply with the Pennsylvania Department of Environmental Protection's request to eliminate the overflows and the authorities' need to regain capacity for future development, both municipalities have adopted a policy of "total sewer rehabilitation" by mini-basin as a part of their RDII reduction programs.

Total rehabilitation by mini-basin means that all sewer system components—including main line, manholes, service laterals, and building sewers located in a mini-basin—are repaired so they meet the same acceptance testing standards as new sewers. House inspections to identify and remove suspect inflow sources are also performed at every property in the mini-basin. System components are investigated and repaired during dry weather conditions. WW conditions are used to assess RDII removal and verify repairs.

Since a mini-basin generally comprises 100–300 connections, agreements between property owners and each authority for investigation and repair of service laterals and building sewers are required. Both the Susquehanna Township Authority and Lower Paxton Township Authority now have policies where observation tees are installed at every property in the mini-basin, and all service laterals and building sewers are air tested and repaired or replaced until they pass an air test. This work is completed at the expense of each authority and not the property owner.

Starting in 1997, a pilot study of the above approach occurred in a single mini-basin. Pre-rehabilitation flow monitoring indicated an average peak RDII contribution of 8,900 gallons per day (gpd) from each of the 113 house connections in the basin. Rehabilitation proceeded in two phases. In the first phase, television inspections were used to identify 17 leaking service laterals. These were excavated at the property line and subjected to an air test. Only one building sewer (typically cast iron) failed the test, and seven service laterals (typically asbestos cement pipe) failed. It was surmised that the failures were at the connection of the building sewer and service lateral. These failed laterals were replaced or lined, all manholes in the basin were rehabilitated, and cured-in-place sectional liners were used to rehabilitate defects in the main line sewers.

Subsequent to the first phase rehabilitation, television inspection indicated that an additional 11 laterals were leaking, which had not been found to leak during previous inspections. The migration of groundwater to these laterals was suspected to be the cause. Comparison of peak flows with those in control basins suggested a 50% reduction in peak hourly RDII had been achieved. It was believed that better results could be achieved if the groundwater migration issue could be addressed by a comprehensive rehabilitation.

The second phase of the project included lining or replacing all service laterals and building sewers that failed an air test (36 and 10, respectively). In addition, sump pumps were disconnected from the sanitary sewer at 15 residences. Based on observed flows from the mini-basin that were derived from the subtraction of flows from three flow monitors, the conclusion was that substantial elimination of RDII had been achieved. The cost for this pilot study was approximately $500,000 (Hilderhoff and Wendle, 2001).

2.1.3.18 Lot Drainage Characteristic Study, City of Edmonton, Alberta

Foundation drain systems around residential basements contribute significant RDII into the sanitary or combined sewers to which they are connected. The additional RDII can cause sewer overloading, an increase in sewer overflows, and additional capital and operating costs for the collections system and treatment plant. In 1990, a test facility was constructed in Strathcona County to 1) determine to what extent different lot drainage characteristics affect foundation drain flows and 2) recommend general improvements to design and construction practices to reduce these flows. The facility's foundation drains, sump pump installations, and site instrumentation were designed to carry out lot drainage characteristics testing.

Soil type was found to be the most significant factor impacting foundation drain responses. Silty soils produce the lowest responses, followed by clay and sandy soil. The rank of the significance of the other variables, in descending order, is degree of compaction, backfill zone grading, and roof leader location.

Some of the recommendations made from the study results were to:

- Consider using silty soils as the preferred backfill material. The flow responses measured in uncompacted clay and sand were 20 times higher than with silty backfill.
- Compact the backfill for new construction. Compaction of clay backfill resulted in a 90% lower flow response. Compaction of silt soils resulted in a 50% lower flow response. Compaction did not affect response in sandy soils.
- Provide engineered backfill where sand must be used, including installation of an impermeable synthetic barrier.
- Provide grading with a 5–10% slope draining away from the house for at least 5′ beyond the limit of excavation. This had the greatest impact for sandy soils, resulting in about a 50% reduction in response compared with adverse grading.
- Minimize the width of the construction zone—to maintain the less permeable properties of surrounding undisturbed soil and to reduce the potential for settling.
- Extend roof leaders beyond the backfill zone and at least 5′ from the foundation wall. The effect was small for sandy soil and for compacted clay and silt. For uncompacted clay and silt, however, this action reduced peak flow response by about 30%. It was also noted that discharge of the roof leader next to the house in uncompacted clay soil accounted for 60% of the measured flow response (CH2M Hill, 1994).

2.1.3.19 Huntington Ridge Area Wye Grouting Project, Allegheny County, Pennsylvania

South Fayette Township Municipal Authority (SFMA) serves the Township of South Fayette located in the southwest portion of Allegheny County, Pennsylvania. The system serves a primarily residential suburban community and comprises just over 100 miles of sanitary sewer and four pump stations. As a condition to connection to the Allegheny County Sanitary Authority system, SFMA entered into an agreement specifying penalties for excess I/I. In 1997, SFMA contracted for the televising of more than 50,000 LF of sewer.

The inspection indicated that of 805 wyes, an estimated 639 appeared to have slightly open joints at the point of connection to the lateral. Approximately 500 wyes were repaired by grouting. Pre- and post-rehabilitation flow data was analyzed to determine how much flow reduction was achieved. Although the analysis proved to be inconclusive as to whether the project had any significant impact on peak flow reduction during precipitation events, it clearly demonstrated that there was a beneficial impact on base infiltration.

2.1.4 Modeling Sewer Systems and RDII Reduction

Modeling of sewer flow and quality is essential in simulating existing conditions (real-time control) and minimizing sewer overflows. Modeling has also provided an increasingly reliable means to test alternatives and controls before implementation in the field (Zheng and Novotny, 1991). As a result, modeling has been useful in analyzing problems in both combined and sanitary sewers, including sewer overflow effects on receiving water quality, flooding and hydraulic problems, RDII problems, and rehabilitation effect assessment (*Existing Sewer Evaluation and Rehabilitation—WEF MOP*, 1994.

A good sanitary sewer system model that accurately depicts RDII has many benefits. Swarner and Thompson (1994) list the following uses for the runoff/transport model in the King County Department of Natural Resources, Seattle, Washington:

- Determine peak flows of various return intervals
- Plan and design conveyance and treatment facilities based on peak flow estimates
- Quantify RDII problems and differences in service basins
- Estimate rates of RDII from direct I/I
- Compare relative RDII contributions from different basins
- Determine where additional flow monitoring would be most effective
- Estimate the impact of RDII control options
- Evaluate the effectiveness of sewer rehabilitation projects

The most significant problem in modeling sewer systems with respect to RDII is to accurately depict infiltration. Direct inflow is relatively easy to model because it is apparent with any storm, and many storms can provide good data that can be used for calibration (Swarner and Thompson, 1994). Infiltration, however, can be difficult to model because it generally comes from unknown sources, through unknown ground conditions, and into pipes of questionable condition. As a result, good flow and rainfall data is of paramount importance to any meaningful effort to accurately model infiltration (Swarner and Thompson, 1994).

WERF (Bennett et al. 1999) analyzed and compared the methods to quantify and simulate RDII. The following information supplements the WERF report. In recent years, several notable methods have been used to model sewer systems and RDII:

- ***Runoff/Transport Model.*** Uses a variant of the runoff and transport blocks of the USEPA Stormwater Management Model (SWMM) to simulate stormwater runoff using a kinematic wave approximation for overland and minor sewer flow, whereby initial rainfall is removed before excess rainfall begins to contribute to sewer flow (Vitasovic et al. 1990). This method has been adapted for use in sanitary sewers and simulates base flow and runoff from impervious and pervious areas into the sewer system (Swarner and Thompson, 1994).
- ***RDII Source Simulation Modeling.*** Simulates RDII sources in a system that is not surveyed extensively. This method, however, makes generalizations based on physical surveys and inspections (Gray and Nogaj, 1990).
- ***Applying Geographic Information System Technology to Sewer System Modeling.*** Creates a hydraulic simulation model that demonstrates the effectiveness of flow equalization storage to reduce overflows. The method also provides design information for conceptual designs of proposed facilities, including pump stations, pipelines, and storage basins (Miles et al. 1996).

- ***Stochastic Modeling.*** Simulates rainfall sewer flow using a two-part transfer function model that accounts for the transfer of input rainfall into sewer RDII flow and dry weather sewage contributions and disturbances (Zheng and Novotny, 1991).
- ***U.S. EPA SWMM.*** Simulates storm-induced RDII in eight sanitary sewer service basins and directly evaluates the impact of RDII flow, while helping to formulate alternatives for dealing with excess RDII flow (Loucks and Johnson, 1991).
- ***Hydrologic Modeling for Excessive RDII.*** Simulates stormwater inflow and rainfall-dependent infiltration using three tank models based on the characteristics of sewer pipes and drainage areas (Wada and Miura, 1997).
- ***Distributed Physically Based Model.*** Describes the surrounding geohydrological processes and their interaction with a sewer network, similar to how dynamic pipe flow modeling describes the hydraulics of a system. The goal is to be able to analyze the effects of future changes in the geohydrological system (Gustafsson et al. 1996; Gustafsson et al. 1997).
- ***Antecedent Rainfall.*** Correlates rainfall occurring over various periods in the past with RDII flows at any given hour. Antecedent precipitation index (API) values are created for periods of several hours to six months and correlated with flows at any given hour. The time period of the API is used to separate total RDII flows into long-term groundwater infiltration (GWI), short-term infiltration resulting from local buildup of soil moisture, and direct and indirect inflow (Merrill and Butler, 1994).

2.1.5 Predicting RDII Reduction Following Sanitary Sewer Rehabilitation

The value of evaluating and predicting the cost-effectiveness of sewer rehabilitation projects is of particular importance, as many programs have demonstrated a disparity between anticipated and actual RDII reduction (Fernandez, 1986). It is not uncommon for utilities to develop optimization and cost-benefit models to evaluate rehabilitation alternatives with respect to overall cost (DeMonsabert and Thornton, 1997; Dorn and Miles, 1997). Few utilities, however, have developed models that take into account predicted RDII reduction as a result of rehabilitation alternatives. The predictive tools that utilities have developed for measuring RDII reduction following rehabilitation have, by and large, been specific to particular projects and cannot be universally applied (Kurz, 1997; Swarner and Klusman, 1995).

Even so, the results of rehabilitation studies are often not reported. If they are, they typically do not include enough standard information to be instructive about the project to other communities. Keefe (1998) suggests that with implementation of a pre- and post-rehabilitation flow measurement standard, communities could learn from each other and affect a global improvement in RDII reduction in sanitary sewer rehabilitation projects.

2.2 Phase I—Data Collection

Phase I of the literature review involved surveying agencies to identify projects that have sufficient information to test RDII rehabilitation methods and to predict the effect of sewer rehabilitation on peak and average RDII flow. To warrant further investigation, projects must meet the tests of good data gathering and comprehensive sewer system rehabilitation.

Phase I includes an identification of agencies that have measured I/I before and after sanitary sewer rehabilitation projects. Specific projects were identified that have sufficient information to test methodologies that predict the effect of sewer rehabilitation on peak and average RDII flows. To test the methodologies, the projects must have high-quality data and entail comprehensive sewer system rehabilitation.

The University of Washington (UW) conducted a literature search to identify documents related to RDII reduction following sanitary sewer rehabilitation. The online catalogs Compendex and Water Resources Abstracts were searched using the following keywords:

- Flow estimation/measurement
- Flow optimization/prediction/projection
- Hydrologic models/modeling
- Infiltration
- I/I
- Inflow
- Rainfall and sewers
- Runoff
- Sanitary flow estimation
- Sanitary sewers
- Sewage flow
- Water infiltration
- Water I/I
- Water inflow
- Watershed models
- WW
- WWF projection

The searches identified more than 1,200 citations having these keywords. These citations were screened for relevance to the project, with 63 being identified as of particular relevance. All the literature was eventually compiled in a database.

2.3 Phase II—Data Screening

To best serve the greatest number of WERF subscribers, it was desirable to find projects from each climatic zone identified by the National Climatic Data Center (NCDC), a division of the National Oceanic and Atmospheric Administration (NOAA). As local weather conditions can affect the frequency, duration, and magnitude of rainstorms, these conditions have a similar effect on RDII flows.

Figure 2-1 shows the climatic zones of the United States as described by NCDC.

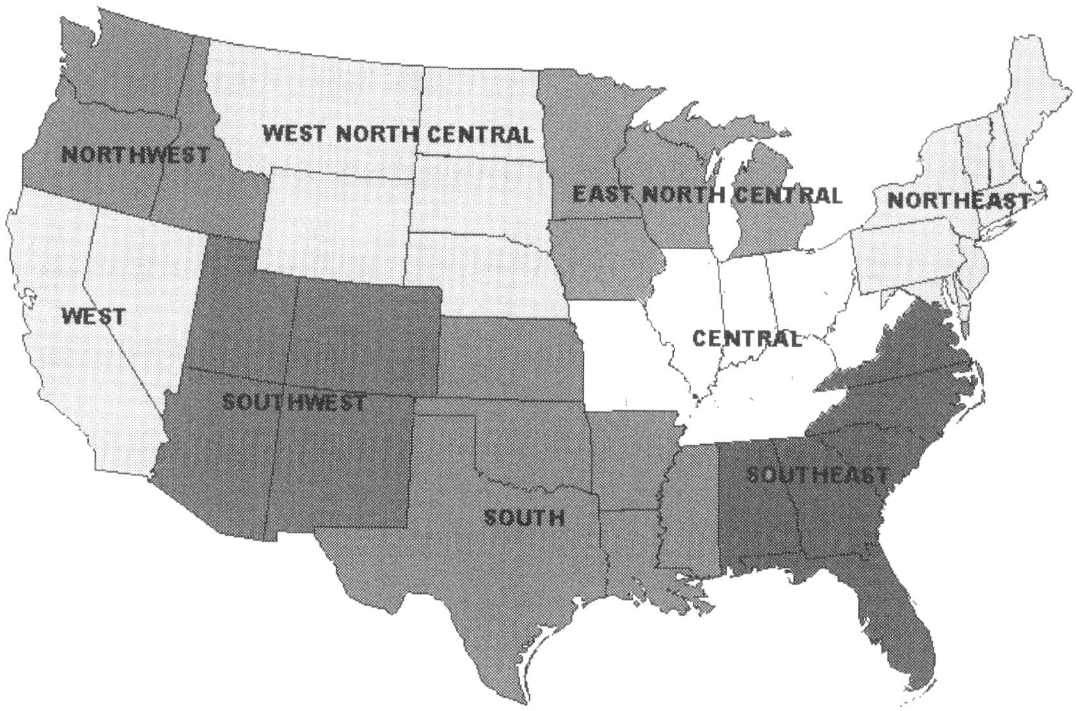

CLIMATE PREDICTION CENTER, NOAA

Figure 2-1. Climatic Zones of the United States.

UW and Brown and Caldwell engineers identified 56 utilities as potential candidates for well-documented RDII removal rehabilitation projects. In all, 44 utilities were contacted, 28 of which either had completed preliminary screening questionnaires or had supplied enough anecdotal information to qualify for further consideration in this study.

Table 2-1 presents the agencies that appeared to have promising RDII projects to evaluate in Phase III.

Table 2-1. Candidate Utilities for Well-Documented RDII Removal Rehabilitation Projects.

Entity/Utility (number of projects under consideration)	*City/Area*	*State*	*Climatic Region*
CMUD (4)	Charlotte	NC	Southeast
Lacey, Olympia, Tumwater, and Thurston (LOTT) (2)	Olympia	WA	Northwest
City of San Luis Obispo (2)	San Luis Obispo	CA	Southwest
City and County of Honolulu (2)	Honolulu	HI	Pacific
East Bay Municipal Utilities Department	Oakland	CA	West
King County Department of Natural Resources	Issaquah	WA	Northwest
King County Department of Natural Resources	Kent	WA	Northwest
Milwaukee Metropolitan Sewerage District (MMSD)	Milwaukee	WI	East North Central
Nashville Metropolitan Water Services (12)	Nashville	TN	South
Stege Sanitary District	El Cerrito	CA	West
Sydney Water Corporation (4)	Sydney		Southeast Australia
Clean Water Services[a] (4)	Hillsboro	OR	Northwest
MMSD	Oak Creek	WI	East North Central
McCandless Township Sanitary Authority (3)	Pittsburgh	PA	Central
Susquehanna and Lower Paxton Townships	Harrisburg	PA	Central

[a]formerly USA of Washington County.

From the table, it is evident that not many communities have gathered sufficient data for the project's objectives. In particular, there is a lack of projects to study in the west north central and central climatic regions.

2.4 Phase III—Project Evaluation

For reviewing the short-listed projects in more detail, the project team and members of the WERF Project Review Committee developed a comprehensive list of data to be collected in the following areas:

- System conditions
- Project information
- Evaluation methodology
- Rainfall records
- Data compilation method used
- RDII quantification method used

Based on this input, the project team developed a questionnaire and companion database to ensure consistent data gathering and the ability to quickly access, sort, and summarize the information.

On review of the followup information, specific projects were selected for further requests for information and review. Efforts to acquire detailed data for analysis proved unsuccessful for several utilities listed in Table 2-1. For the most part, information had not been archived in a manner suitable for use in this study. Table 2-2, as a result, lists the projects selected for detailed analysis.

Table 2-2. Selected Utilities for Well-Documented RDII Removal Rehabilitation Projects.

Entity/Utility (number of projects reviewed)	*City/Area*	*State*	*Climatic Region*	*Selected as Case Study*
LOTT (2)	Olympia	WA	Northwest	Yes
City and County of Honolulu (2)	Honolulu	HI	Pacific	Yes
King County Department of Natural Resources	Issaquah	WA	Northwest	Yes
King County Department of Natural Resources	Kent	WA	Northwest	Yes
MMSD	Milwaukee	WI	North Central	Yes
Clean Water Services[a] (3)	Hillsboro	OR	Northwest	No
MMSD	Oak Creek	WI	North Central	Yes
McCandless Township Sanitary Authority (3)	Pittsburgh	PA	Central	Yes
Susquehanna and Lower Paxton Townships	Harrisburg	PA	Central	No

[a]Formerly USA of Washington County.

On receipt of complete information, the project team reviewed the completeness, adequacy, and perceived accuracy of the flow and rainfall data. As a result of this review, three projects were not selected as case studies to present in this report.

CHAPTER 3.0

RDII Quantification Methods

Municipal engineers and consultants have historically used a variety of methods to quantify RDII. The differences among these methods make comparisons of RDII removal between different projects difficult. The purpose of Chapter 3.0, therefore, is to compare the widely used methods and establish guidelines on the proper quantification of RDII for the case studies presented in this report.

3.1 Characterization of RDII

As SSO regulations become more stringent, the evaluation of RDII removal in separated sewer systems is becoming increasingly important. SSO is often caused by RDII from a storm event. Increased flow from streets, buildings, public and private property, and surface water enters the sanitary sewer system through cracked pipes, manholes, or illegal connections.

Figure 3-1 illustrates the typical sources of RDII.

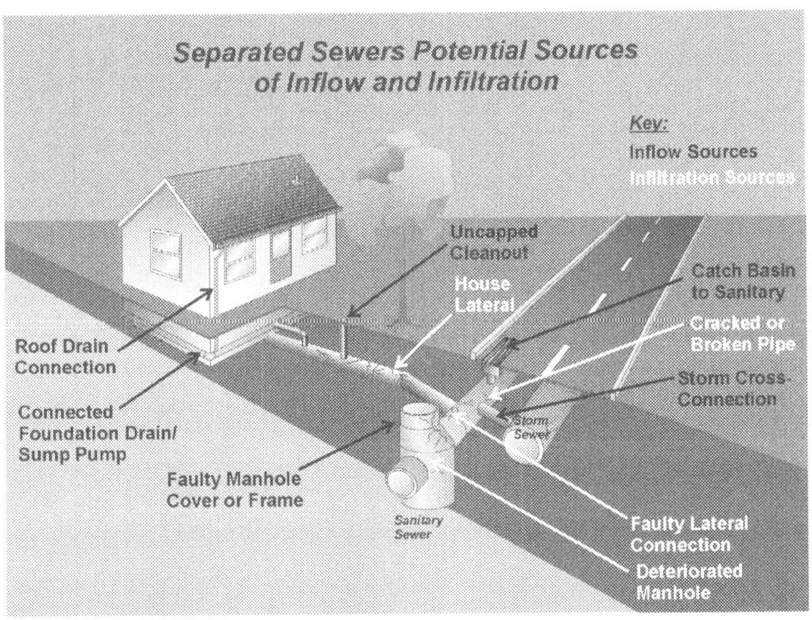

Figure 3-1. Typical Sources of I/I.

Collection and treatment systems are exceeding design capacities well ahead of their original design life because of excessive RDII, resulting in surcharging, backups, bypasses, and reduced treatment efficiency. As a result, operation and maintenance (O&M) costs increase since greater-than-designed flows result in excessive wear, higher power demands, and operation outside the optimum range of efficiency. The new regulations requiring reduction in sewer overflow events and more extensive wastewater treatment create additional costs for a community. By reducing RDII in a system, however, SSO can be alleviated and flow to the treatment facility decreased.

In 1999, WERF conducted research to help municipal agencies quantify and simulate RDII. This chapter summarizes information contained in the document produced by this research effort, *Sanitary Sewer Overflow (SSO) Flow Prediction Technologies* (WERF Project 97-CTS-8 Final Report, 1999). RDII—the additional flow in the sewer system from rainfall or snowmelt—is defined as the portion of a sewer flow hydrograph above the normal dry weather flow pattern.

Figure 3-2 shows RDII flow in a sanitary sewer system. Note the increased flow (RDII) caused by rainfall in the sewershed. RDII contributes to SSOs and increased flows to the wastewater treatment facility.

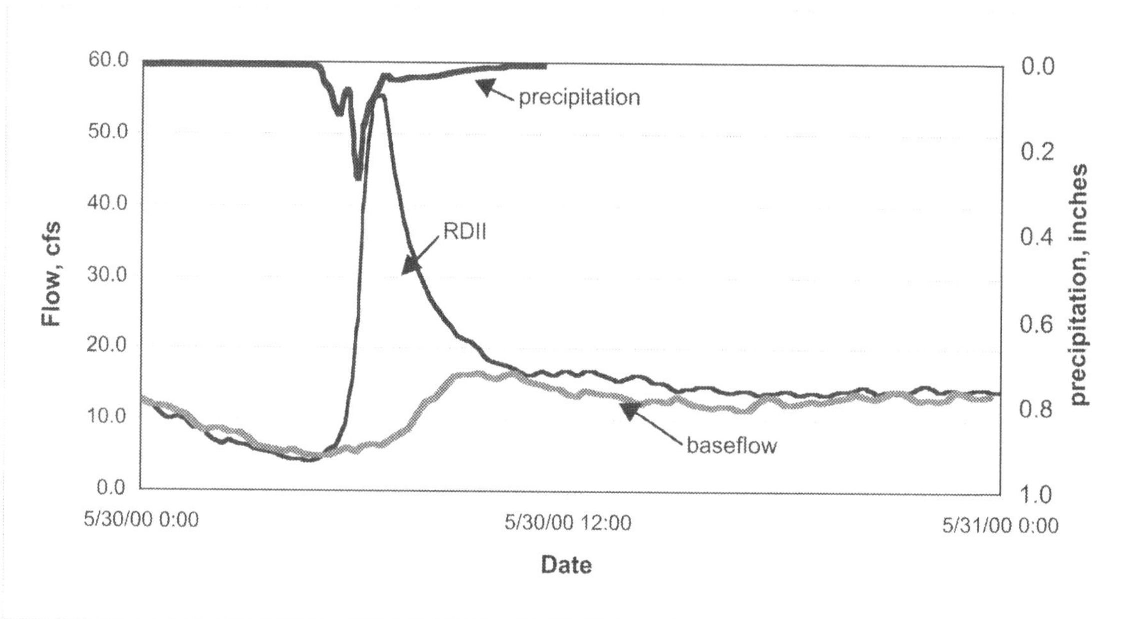

Figure 3-2. Typical RDII and Base Flow Hydrograph.

Research conducted in WERF Project 97-CTS-8 was designed to produce a guide for agencies and consultants in selecting and using appropriate RDII flow projection methodologies. Eight methods to quantify RDII were summarized and recommendations were made on the applicability of the methods to a variety of situations evaluated.

The WERF project categorized the following identified RDII hydrograph generation methods:

- Constant unit rate methods
- Percentage of rainfall volume (R-value) methods
- Percentage of streamflow methods
- Synthetic unit hydrograph methods
- Probabilistic methods
- Predictive equation based on synthetic rainfall/flow regression methods
- Predictive equation based on synthetic streamflow and basin characteristics methods
- RDII as a component of hydraulic software methods

Table 3-1 summarizes all eight methods, including the advantages and disadvantages of each and where the method was applied. Sections 3.2 through 3.9 describe each method in detail.

Table 3-1. Summary of RDII Hydrograph Generation[a].

Method	*Summary*	*Advantages*	*Disadvantages*	*Applications*
Constant unit rate	Constant unit rate (e.g., gallons per inch rainfall per acre, gallons per acre per land use, gallons per inch rainfall per capita) based on sewershed characteristics.	Easy to apply.	Unit rates developed in one area may not pertain to another. Does not represent temporal variation.	St. Paul, MN Edmonton, Alberta Corvallis, OR Salem, OR Milwaukee, WI
Percentage of rainfall volume (R-value)	Ratio between volume of RDII at a monitoring location and rainfall volume falling on area served by monitor. Once enough data points are established, a relationship can be estimated.	Develops range of RDII volumes for monitored sewersheds.	Does not represent temporal variations. Requires data from several representative storms.	Dallas, TX Cobb Co., GA Montgomery, AL Sepulpa, OK ADS Slicer©
Percentage of streamflow	Relationship is developed between sewer flow-monitoring data and streamflow data.	Easy to apply if streamflow data is available.	Sewer flow is usually affected by a much smaller watershed than that reflected in available streamflow records.	Tested for: Minneapolis, MN Milwaukee, WI
Synthetic unit hydrograph	Assumes that RDII responds to rainfall volume and duration in the same manner as stormwater runoff and the shape of an RDII hydrograph is a function of basin characteristics.	Can be scaled in proportion to rainfall and duration for a range of storms once unit hydrograph is developed.	May require several unit hydrographs to represent components of an RDII hydrograph.	Edmonton, Alberta Montgomery, AL
Probabilistic	Frequency analysis of peak RDII flows, which results in a relationship between peak RDII flow and recurrence interval.	Gives good representation of probability of occurrence of peak RDII flow.	Does not provide RDII hydrograph for specific precipitation events.	Tested for: Minneapolis, MN

Table 3-1. Summary of RDII Hydrograph Generation[a] (continued).

Method	*Summary*	*Advantages*	*Disadvantages*	*Applications*
Predictive equations based on rainfall/ flow regression	Uses multiple linear regression methods to derive a relationship between hourly rainfall and RDII.	Generates highly representative predictive equations for storm events and antecedent rainfall.	Sensitive to anomalies in flow and rainfall data used to generate equations. May need several equations to represent storms of different character.	Alexandria, VA Corvallis, OR Salem, OR Portland, OR Albany, OR Eugene/ Springfield, OR Pawtucket, RI Narragansett Bay, RI
Predictive equations based on synthetic streamflow and basin characteristics	Uses multiple regression to correlate outputs from a calibrated continuous hydrologic simulation model of a watershed (daily subsurface and surface flow) to sewer flow components (daily GWI and RDII).	Applies hydrologic theory to the generation of RDII. Well-calibrated correlation will give good results.	Usually only justified if calibrated hydrologic model is already available. Otherwise, method may be overly labor intensive.	Milwaukee, WI
RDII as component of hydraulic software	Various hydrology/ hydraulic packages include methods for generating RDII.	Easy to apply and readily available as part of software.	Must own software to use. User usually has minimal options for customizing to individual applications.	European Union

[a]From *Sanitary Sewer Overflow (SSO) Flow Prediction Technologies*, WERF Project 97-CTS-8 (1999).

3.2 Constant Unit Rate Method

For the constant unit rate method, the assumption is that RDII can be estimated as a constant unit rate based on sewershed characteristics. RDII from large areas should be based on independent variables such as area (gallons per acre), land use (gallons per acre per land use), population (gallons per capita), and pipe length (gallons per inch-diameter-mile (idm)). For small areas, independent variables include characteristics related to known system defects, pipe material and age, roof drain connections, backfill material, and compaction.

A limitation of this method is that the constant unit rates developed for one system may not apply well for other systems. In addition, unit rates developed in spring may not be applicable in summer. The recommendation, therefore, is to assess the validity of any constant unit rate for each system and moisture condition. This can be done quantitatively by using flow data or qualitatively by evaluating system land uses and characteristics.

Table 3-2 summarizes evaluation criteria for the constant unit rate method.

Table 3-2. Constant Unit Rate Method Data Summary[a].

Evaluation Criteria	*Discussion*
Data requirements	Characterization of sewershed tributary to flow data location by one or more of the following parameters: area, land use, population, diameter/length of pipe, age of pipe.
Quantity of data required	Low: requires fewer data items than other methods.
Precision of data required	Gross: data must be estimated only and applied consistently in developing and applying unit rates.
Form	Constant rate of RDII flow per parameter unit per time.
Accuracy	Low: requires data less accurate than that mandated by other methods.
Application	Quick analysis of sewerage system capacity relative to hypothetical WW peak flows.

[a]From *Sanitary Sewer Overflow (SSO) Flow Prediction Technologies,* WERF Project 97-CTS-8 (1999).

3.3 Percentage of Rainfall Volume (R-Value) Method

The R-value method draws a relationship between RDII and rainfall volumes. The ratio between the volume of RDII at a monitoring location and the volume of rainfall that falls on the drainage area served by the monitor is defined as the R-value. To compute the R-value, average base flow is first determined by analyzing metered data for extended dry periods. WW RDII is then determined by subtracting the average base flow component from metered data during wet periods. The volume of RDII for significant storm events is calculated and compared to rainfall data to tabulate R-values for individual storms. Because this method results in the volume of RDII only, it is not useful for predicting peak rates of flow.

Table 3-3 summarizes evaluation criteria for the percentage of rainfall volume (R-value) method.

Table 3-3. Percentage of Rainfall Volume (R-Value) Method Data Summary[a].

Evaluation Criteria	*Discussion*
Data requirements	One or more rainfall gauges reasonably representative of monitoring sewersheds; multiple flow-monitoring/rainfall events.
Quantity of data required	High: requires flow and precipitation data for many storms.
Precision of data required	Intermediate: requires accurate representation of event volumes, but not necessarily of intra-event variations.
Form	Single factor relating rainfall/snowmelt to RDII volume.
Accuracy	Medium: requires less accurate data than methods relying on subtle variations between observations to establish trends used in the method.
Application	Representation of RDII storm response volume.

[a]From *Sanitary Sewer Overflow (SSO) Flow Prediction Technologies,* WERF Project 97-CTS-8 (1999).

3.4 Percentage of Streamflow Method

In certain circumstances, streamflow can reliably predict RDII because the combinations of hydrologic conditions that increase streamflow are similar to those that affect RDII entering a sewer system. A relationship can be developed between sewer flow-monitoring data and streamflow data. In applying this relationship, the assumption is that the sewer system conditions are consistent with the conditions in the monitored sewershed areas.

This method is not widely used, its major limitation being that sewersheds are typically much smaller than gauged watersheds. Sewershed flow, therefore, is affected by different rainfall patterns than watersheds. In addition, the time of concentrations is different. During large events, the sewer may cease to admit RDII once they surcharge, while the stream would continue to admit runoff. In addition, peak streamflow may be attenuated by storage via overbank flooding that has no direct corollary in the sewer system. Overbank areas also typically have higher friction factors than the normal channel, again attenuating larger peak flows different than that experienced in sewers.

Table 3-4 summarizes evaluation criteria for the percentage of streamflow method.

Table 3-4. Percentage of Streamflow Method Data Summary[a].

Evaluation Criteria	*Discussion*
Data requirements	Stream gauge data corresponding temporally with sewerage system data.
Quantity of data required	High: requires long-term streamflow data in a watershed analogous to the sewershed.
Precision of data required	Gross: requires the monitoring of complete watersheds and sewersheds only.
Form	Single factor that relates measured streamflow to measurements of RDII.
Accuracy	Medium: requires less accurate data than methods relying on subtle variations between observations to establish trends used in the method.
Application	Representation of RDII storm response volume.

[a]From *Sanitary Sewer Overflow (SSO) Flow Prediction Technologies,* WERF Project 97-CTS-8 (1999).

3.5 Synthetic Unit Hydrograph Method

The synthetic unit hydrograph method uses the unit hydrograph theory. The assumption is that RDII responds to rainfall volume and duration in a consistent manner and that the shape of an RDII hydrograph is a function of basin characteristics. It assumes that the time duration of the RDII hydrographs attributable to rainfall of a specific duration is constant and that the peak flow is proportional to the volume of rainfall. On the basis of these assumptions, unit hydrographs for RDII can be developed and calibrated to the monitoring data.

A limitation of this method is that the accuracy of the predictions is directly related to the accuracy of the rainfall data. This is because the unit hydrograph flow estimates are generated from the discrete rainfall measured within or near each monitored basin. Poor rainfall data will limit the accuracy of this method and any other rainfall-derived RDII quantification method.

Table 3-5 summarizes evaluation criteria for the synthetic unit hydrograph method.

Table 3-5. Synthetic Unit Hydrograph Method Data Summary[a].

Evaluation Criteria	*Discussion*
Data requirements	Stormwater runoff parameters in sewershed area; several sets of precipitation and corresponding sewerage system flow response data.
Quantity of data required	Medium: requires several storms but does not demand continuous data.
Precision of data required	Intermediate: data must represent only the general shape of hydrographs and hyetographs.
Form	RDII response hydrograph calculated from a hyetograph.
Accuracy	High: if well calibrated.
Application	Full representation of RDII hydrograph for system characterization and input to hydraulic models.

[a]From *Sanitary Sewer Overflow (SSO) Flow Prediction Technologies*, WERF Project 97-CTS-8 (1999).

3.6 Probabilistic Method

The probabilistic method is a frequency analysis of peak RDII flows. This method is described in most hydrology textbooks and can be applied to flow rates or volumes. To apply the method, the peak value (rate or volume) is identified for each month in a period of record that covers a representative period of wet and dry season. A statistical analysis is performed to determine recurrence intervals for each value. The peak values are ranked and plotted on log-probability paper. Values for various recurrence intervals can be determined from the graph.

The product of a probabilistic analysis is a relationship of peak RDII flow to the recurrence interval. For example, the probabilistic method would specify the magnitude of a peak RDII flow that has a 4% probability of occurring (25-year return period).

The major limitation of the probabilistic method is that it requires long-term data. Another limitation is that its application is limited to the conditions observed.

Table 3-6 summarizes evaluation criteria for the probabilistic method.

Table 3-6. Probabilistic Method Data Summary[a].

Evaluation Criteria	*Discussion*
Data requirements	Extensive series of continuous flow records.
Quantity of data required	High: requires consistent sewer flow data for all peaks for a prolonged period of time.
Precision of data required	Gross: requires the monitoring of complete sewersheds only.
Form	Monthly recurrence interval of RDII peak flows.
Accuracy	Gross: relies directly on sewer flow data and, therefore, is not subject to compounding errors that occur in extrapolations.
Application	Characterizing RDII peak flows according to recurrence interval.

[a]From *Sanitary Sewer Overflow (SSO) Flow Prediction Technologies,* WERF Project 97-CTS-8 (1999).

3.7 Predictive Equations Based on Rainfall/Flow Regression Method

The rainfall/flow multiple regression method uses rainfall and flow-monitoring data to derive a relationship between rainfall and RDII. This relationship, expressed as an equation, is then used to estimate RDII flows for the selected rainfall periods. After obtaining a satisfactory fit between measured and estimated flows, the equation can be used to characterize, as well as quantify, RDII flows. This method also assumes that the basin flows would behave according to the regression in rainfall conditions outside those observed.

The method assumes that a multiple linear relationship exists between past rainfall and RDII. The following steps are taken to develop predictive RDII relationships for each flow-monitoring location (the first three steps are common to most RDII quantification methods):

1. Flow data is separated into dry and wet periods according to the rainfall record.
2. A typical dry weather 24-hour (diurnal) base sanitary flow pattern is established.
3. Flow attributable to RDII is separated out by subtracting the dry weather hydrograph from the WW hydrograph for a particular rainfall event.
4. Rainfall data from 15–60 days preceding each hourly flow measurement is summed for incremental periods (e.g., 1 hour, 2–3 hours, 4–6 hours, 7–12 hours, 12–24 hours, 1–2 days, 4–7 days, and 7–15 days).
5. Multiple linear regression is performed to identify the relationship between the rainfall sums and measured flow. It determines the regression coefficients (Ci) for each rainfall sum. Regression equations are in the form of:

   $$\text{RDII} = \text{C1} \times \text{Rain}_{\text{1-hour}} + \text{C2} \times \text{Rain}_{\text{2-hour}} + \ldots\ldots\ldots\ldots \text{C8} \times \text{Rain}_{\text{7–15 days}}$$

   $$\text{Total Flow} = \text{Base Sanitary Flow} + \text{RDII}$$

   Ci derive from a multiple linear regression of RDII flow against rainfall summed over the various time periods.
6. Other periods of rainfall are then applied to the created regression equation to verify the simulated RDII against actual flow data.

The regression models can be adjusted by manipulating the coefficients of the regression to improve the fit between predicted and actual flow.

Table 3-7 summarizes evaluation criteria for the predictive equations based on the rainfall/flow regression method.

Table 3-7. Predictive Equations Based on Rainfall/Flow Regression Method Data Summary[a].

Evaluation Criteria	*Discussion*
Data requirements	Extended wet season flow and precipitation data. Rainfall gauges should reasonably represent rainfall in drainage area of flow monitors.
Quantity of data required	Medium.
Precision of data required	Detailed.
Form	RDII response hydrograph.
Accuracy	High: if application storms are similar to storms for which predictive equations are generated.
Application	Full representation of RDII hydrograph for system characterization and input to hydraulic models.

[a]From *Sanitary Sewer Overflow (SSO) Flow Prediction Technologies,* WERF Project 97-CTS-8 (1999).

3.8 Predictive Equations Based on Synthetic Streamflow and Basin Characteristics Method

Hydrologic theory includes several methods that adequately predict basin surface and subsurface runoff and streamflow responses to rainfall. These methods can be extrapolated to predict RDII by correlating the hydrologic responses to sewer flow response.

For this method, a continuous hydrologic simulation program is calibrated to long-term streamflow records. The hydrologic calibration is adapted to gauged sewersheds by developing a multiple linear regression between the sewer flow components (daily GWI and RDII) and hydrologic model outputs (daily subsurface runoff and daily surface runoff). A second regression analysis can be used to extrapolate RDII flows in the study basin to other basins within a service area. This provides a method to estimate RDII flows for each basin by using physical characteristics such as the following:

- Total sewer length
- Number of manholes
- In-miles of sewer
- Residential, industrial, and commercial area
- Number of residential, industrial, commercial, and institutional buildings

The sewershed for which RDII is being generated must not necessarily be encompassed by the gauged watershed; however, it should be of similar geohydrologic classification. In addition, the gauged watershed should be similar in dimension and size to the sewershed, although some applications have been successful when the watershed has been substantially larger.

RDII predictions based on streamflow are limited to those areas where streamflow records are available for small watersheds in proximity to the sewershed. Where the appropriate data is available, its use is often limited by the large data manipulation required.

Table 3-8 summarizes evaluation criteria for the predictive equations based on the synthetic streamflow and basin characteristics method.

Table 3-8. Predictive Equations Based on Synthetic Streamflow and Basin Characteristics Method Data Summary[a].

Evaluation Criteria	*Discussion*
Data requirements	Calibrated continuous hydrologic simulation program; precipitation record; daily sewer flow data.
Quantity of data required	High.
Precision of data required	Intermediate.
Form	Daily RDII volumes in response to rainfall.
Accuracy	Medium.
Application	Representation of RDII storm response volume.

[a]From *Sanitary Sewer Overflow (SSO) Flow Prediction Technologies*, WERF Project 97-CTS-8 (1999).

3.9 RDII as Component of Hydraulic Software Method

Several hydrology/hydraulic software packages include methods for generating RDII. For example, USEPA SWMM®, XP-SWMM®, Mouse®, Hydra®, SewerCAD®, and Hydroworks® include components that predict RDII based on physical parameters entered by the user. Most of the software packages incorporate one or more of the previously discussed RDII quantification methods, although they may not use the same terminology in describing them.

Table 3-9 summarizes evaluation criteria for RDII as a component of the hydraulic software method.

Table 3-9. RDII as Component of Hydraulic Software Method Data Summary[a].

Evaluation Criteria	*Discussion*
Data requirements	Model input parameters.
Quantity of data required	Low to high.
Precision of data required	Varies.
Form	Varies between constant unit rate and temporally varying hydrograph.
Accuracy	Varies according to extent of calibration.
Application	Estimates of hypothetical RDII quantities as a function of model input parameters.

[a]From *Sanitary Sewer Overflow (SSO) Flow Prediction Technologies*, WERF Project 97-CTS-8 (1999).

3.10 Recommended Methods for Quantifying RDII of Case Study Projects

In calculating RDII, the case study projects selected for this report generally utilized one or more of the quantifying methods described in this chapter. To normalize the results of the case studies so that meaningful comparisons could be made for this report, the RDII from each case study was quantified, using the following methods:

- A hydrologic balance model similar to that contained in the SWMM runoff block groundwater routines and the Mouse RDII module (Merrill, 2000) was applied to each case study. The simulation model balances precipitation, evapotranspiration, soil moisture, groundwater, and flow into the sewers. Models were calibrated to measured flow and precipitation for both the pre- and post-rehabilitation periods.
- Calibrated models for pre- and post-rehabilitation periods simulated the hourly flow response to long-term rainfall records. The annual peak estimated RDII flows were then fitted to a statistical distribution to define expected peak RDII flows at various return intervals. Reduction of RDII was computed from the difference in predicted peak RDII flows at specified return intervals. This approach normalizes the inevitable differences in precipitation volumes and patterns between the pre- and post-rehabilitation monitoring periods. Results were checked by reviewing plots of measured RDII versus antecedent rainfall in the previous 24–48 hours (Q versus I plots).
- The methods described above could not be used in one case study, McCandless Township. In that instance, reduction in peak RDII flow was analyzed by comparing measured event peak RDII flows before and after rehabilitation with corresponding events in a nearby control basin.

CHAPTER 4.0

RDII Reduction Evaluation Methods

Once the RDII is quantified and the rehabilitation work is performed to reduce the RDII impacting the sewer system, an RDII reduction evaluation method can be used to estimate the quantity of RDII removed. Chapter 4.0 identifies and discusses several methods for determining RDII removal success for a given project setting. Some evaluation methods discussed can be performed regardless of RDII quantification method used.

Table 4-1 lists several RDII removal projects and the RDII reduction evaluation method used to determine the extent of RDII removal.

Table 4-1. RDII Reduction Evaluation Methods.

Method	*RDII Reduction Evaluation Method*	*Project Author and Year*	*Project Location*
1	Sum of RDII source correction efforts multiplied by unit reduction rates	Western Lake Superior Sanitary District (WLSSD), 1993	Duluth, MN
2	Direct comparison of pre- and post-construction flow-monitoring data	CH2M HILL, 1981	MMSD, Milwaukee, WI
3	Comparison of rehabilitation with control basin R-factors	CDM, 1988	CMUD, Charlotte, NC
4	R-factor correlation analysis with R-factor envelope analysis	Brown and Caldwell, 1998	Honolulu, HI
5	Statistical comparison of continuous simulation models	Merrill and Lukas, 1999	King County, WA[a]
		Davis and Cleveland, 1999	Olympia, WA[a]

[a]Brown and Caldwell report.

Sections 4.1 through 4.5 discuss each reduction evaluation method listed.

4.1 Method 1—Sum of RDII Source Correction Efforts Multiplied by Unit Reduction Rates

In the early 1990s, WLSSD began a program for reducing the RDII in its satellite collection systems. Each community was required to document the RDII source investigation, and

RDII source correction work performed on an annual basis. No specific program was established to determine the exact effectiveness that RDII source correction work was achieving. To determine the result of the efforts, WLSSD agreed to RDII removal credits with its customer agencies. Developed by using best engineering judgment, the values reflect a peak instantaneous rate of RDII flow removal according to the specified RDII removal activity:

- Manhole cover replacement = 30 gallons per minute (gpm) per manhole
- Manhole sealing = 40 gpm per manhole
- Residential foundation drain = 20 gpm per building
- Residential sump pump = 20 gpm per building
- Residential roof drain = 12 gpm per downspout
- Commercial footing drain = 25 gpm per building
- Trunk line replacement = 100 gpm (units undocumented)

To determine the level of success of the program each year, WLSSD multiplied the unit RDII removal factor by the corresponding number of activities.

4.1.1 Advantages

This method has several advantages. First, it is simple to use. Totaling the number of each type of RDII source correction activity performed for the project and multiplying it by the unit rate can obtain RDII removal amounts very quickly. Second, the method requires no pre- or post-rehabilitation flow-monitoring data to estimate the success of the project in terms of amount of RDII flow removed from the system. Post-rehabilitation would be required to determine the residual amount of RDII in the project basin.

4.1.2 Disadvantages

The method provides no actual measure of effectiveness of RDII removal for a project. The unit rates are based on no actual data but on best engineering judgment. The rates would be applied to any type of defect, regardless of whether the defect actually allowed the RDII to enter the system. As a result, the method provides no actual feedback on existing RDII levels, remaining RDII levels, or removed RDII amounts.

4.2 Method 2—Direct Comparison of Pre- and Post-Construction Flow-Monitoring Data

In 1981, the Private Property I/I Pilot Project was conducted for the MMSD. As part of this project, different types of repairs and rehabilitation were performed and analyzed to determine feasibility and associated costs. Repairs performed included:

- Sump pump installation
- Pump-full grouting
- Sewer sausage grouting
- Sliplining

In the areas where foundation drain repairs were made, post-rehabilitation flow data was collected to determine the effectiveness of the rehabilitation. The design storm maximum hour inflow for the contiguous area basin was projected to be 142 gpm and 115 gpm for the pre- and post-rehabilitation monitoring period, respectively. The post-rehabilitation value reflects the inflow reduction due to the disconnection of 13 foundation drains. The average maximum hour inflow removed per sump pump installation was (142 gpm–115 gpm)/13 = 2.1 gpm. Compared with the 2.7 gpm average flow rate based on the individual sump pump monitoring data, the difference represents a 75% effective level of foundation drain inflow removal. Installation of a sump pump with related foundation drain repairs, therefore, was estimated to remove 75% of RDII attributable to the foundation drain.

Lateral repair by sliplining or pump-full grouting was estimated to remove 40% of RDII entering the lateral defects. The RDII removal percentage was calculated by assuming that both rehabilitation methods would remove 50% of RDII entering the lateral between the house and sewer. The portion of the lateral RDII flow was estimated by allocating the projected design storm maximum hour RDII between the lateral and building drain in direct proportion to their lengths. When the estimated rate of RDII removal from the laterals at 50% effectiveness is compared with the total projected maximum hour RDII, the average net removal is 40%.

4.2.1 Advantages

This method provides actual measured results from individual RDII removal activities. The success of the project can be directly measured by comparing pre- and post-rehabilitation data. It is simple to use as it only requires calculating the difference between the maximum flow rates measured during the two monitoring periods.

4.2.2 Disadvantages

Comparing flow-monitoring data from two time periods relies on several assumptions. Differences in groundwater and soil moisture conditions can profoundly affect RDII generation, and this method assumes that the two flow-monitoring periods had similar conditions. Differences in rainfall can influence results on several levels. Rainfall amounts, intensity, and duration all can influence the level of RDII generation. Using this method, a project can actually indicate no RDII was removed if rainfall conditions were substantially "wetter" during the post-rehabilitation monitoring period.

4.3 Method 3—Comparison of Rehabilitation with Control Basin R-Factors

In 1988, CMUD initiated a flow-monitoring program to prioritize areas in the McAlpine Creek, Sugar Creek, and Irwin Creek sewer system for more detailed sewer system evaluation surveys (SSESs) and sewer rehabilitation. Flow data collected in 1988 was used to identify two areas for a pilot SSES and sewer rehabilitation study.

A major goal of the pilot sewer rehabilitation program was to evaluate the effectiveness of different sewer rehabilitation approaches and techniques. For this reason, the project was designed so that entire mini-areas were rehabilitated with the same technique. With this approach, post-rehabilitation flow monitoring allowed an evaluation of the effectiveness of the different rehabilitation techniques.

The analysis methodology used to determine the effectiveness of the rehabilitation was performed in two steps: 1) hydrograph decomposition and 2) linear correlation analysis.

The hydrograph decomposition method involved computing RDII flows from all monitored areas (both rehabilitated and control areas) during the pre- and post-rehabilitation periods. The first step of this analysis was to perform a hydrograph decomposition to determine diurnal dry weather flow hydrographs during the monitoring period and then subtracting the hydrographs from hydrographs measured during a rainfall event. The difference was the flow amount directly resulting from the RDII hydrograph. This procedure determined the volume of RDII resulting from a given rainfall event, which was reported as an R-value (or C coefficient). The R-value is the volume of RDII entering the mini-area. The rainfall volume was computed by multiplying the average rainfall over the mini-area by the sewered area. High R-values indicate mini-areas with high RDII, and low R-values indicate a relatively tight sewer system.

Once the R-values were computed for each rainfall event and flow monitor, the linear correlation analysis was done. The correlation analysis is a method of developing a linear relationship between the R-values of the rehabilitated areas and the R-values of a control area. This linear relationship is established independently, during pre-rehabilitation conditions and post-rehabilitation conditions, by performing a linear regression between the R-values of the rehabilitated monitor and control monitor during each monitoring period. The change in the slope of the best-fit line between the R-values during the pre- and post-rehabilitation monitoring periods is used to indicate the success of the rehabilitation method in reducing RDII in the rehabilitated area.

Figure 4-1 shows the process.

4.3.1 Advantages

This method provides a more complete picture of RDII conditions before and after rehabilitation. The R-factor method, when used for many storms, can provide a statistically meaningful representation of RDII conditions in a flow-monitoring basin. Use of the control basin improves the reliability of the method in "normalizing" the before-and-after construction periods.

4.3.2 Disadvantages

The method only reveals results for the measured flow data available. While use of the control basin can mitigate the impacts of different weather and soil conditions during the two periods, it cannot reliably overcome all conditions. The control basin must be carefully selected so it is similar in size, land use, soil type, groundwater conditions, and system configuration. The method requires at least an entire wet period of data (several months) to reveal reliable conclusions. It is not possible to predict an RDII condition for a standard storm based on the representations developed by this method. The method is based on RDII volume only and does not address changes in peak RDII flow generated in the system.

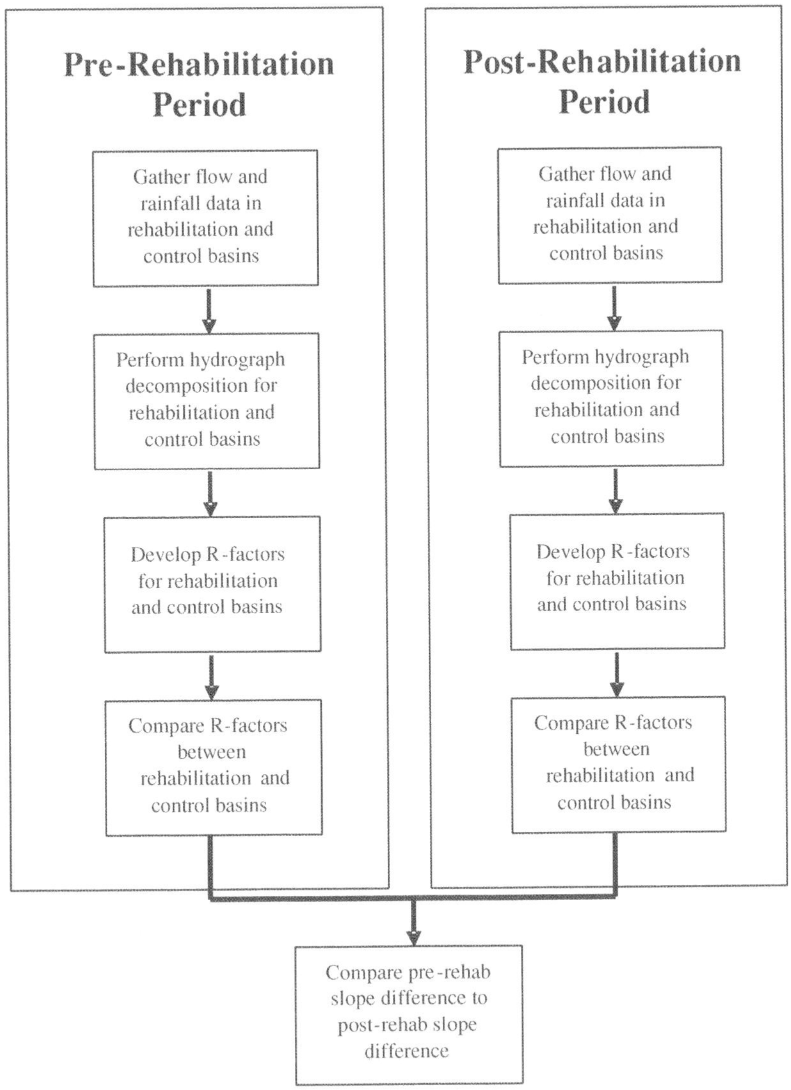

Figure 4-1. Method 3 Approach Schematic.

4.4 Method 4—R-Factor Correlation Analysis with R-Factor Envelope Analysis

The Sewer Rehabilitation and I/I Minimization Pilot Program (RDII Program) was developed for the city of Honolulu. The objective of the RDII Program was to address WW SSOs. Flow and rainfall data was analyzed pre- and post-rehabilitation construction to determine the initial WW RDII characteristics and the extent of WW RDII reduction in the rehabilitated basins. Flow data was collected from flow monitors located at Basin A and Basin B as well as from several control basins that were not rehabilitated to provide a basis for comparison.

The RDII removal effectiveness in the pilot basin was evaluated in two steps: 1) R-factor correlation analysis and 2) R-factor envelope analysis. Pre- and post-rehabilitation R-factors were calculated for both the control and pilot basins for each significant rainfall event. R-factors for the pilot basins were correlated to R-factors for the control basins. The R-factor correlation trend lines for the pre- and post-construction envelopes were developed using the least squares technique. The slopes of the two trend lines were compared to determine the amount of WW RDII reduction.

R-factor correlation plots were completed for each pilot basin monitor, compared with similar plots for each of the three control basin monitors. For each storm event analyzed, R-factors for the selected control basin monitor were plotted on the x-axis and R-factors for the pilot basin were plotted on the y-axis. Each point was designated as either pre or post construction. The slopes of each line were compared to quantify the percent difference between the pre- and post-construction R-factors. This provided a measurement of the R-factor reduction for each pilot basin.

Following R-factor development, the RDII program used the concept of R-factor envelopes to develop design R-factors for flow-monitoring basins. When applied to the rehabilitation effectiveness analysis, an R-factor envelope was drawn for both pre- and post-construction conditions based on the data available. A significant number of storms of varying precipitation volume was analyzed before and after construction to provide sufficient data to approximate both R-factor envelopes. Design R-factor percent reductions were then calculated for both the two-year six-hour and five-year six-hour design storm.

4.4.1 Advantages

This method features the same main advantages as Method 3 since they are initially the same. The method provides a more complete picture of RDII conditions before and after rehabilitation. The R-factor method, when used for many storms, can provide a statistical representation of RDII conditions in a flow-monitoring basin. Use of the control basin improves the reliability of the method in "normalizing" the flow conditions from before and after periods. Adding the R-factor envelope approach improves the ability to determine RDII levels for events other than those specifically measured during the monitoring periods.

4.4.2 Disadvantages

Although use of the control basin can mitigate the impacts of different weather and soil conditions during the two periods, it cannot reliably overcome all conditions. The control basin must be carefully selected so it is similar in size, land use, soil type, groundwater conditions, and system configuration. The method requires at least an entire wet period of data to reveal reliable conclusions. The method is based on RDII volume only and does not address changes in peak RDII flow generated in the system. Storms of varying intensity, but with the same total depth, would not generate different volumes or peak RDII flow under this method.

4.5 Method 5—Statistical Comparison of Continuous Simulation Models

King County, Washington, analyzed pre- and post-construction flow-monitoring and rainfall data of two pilot project areas in the metropolitan Seattle area. Once calibrated on measured flow and rainfall data, long-term rainfall records (40 years or more) were run through the models to develop a database of RDII flow for every hour of the record. This database was then used to develop peak RDII flow occurrence frequency distributions to estimate peak RDII flow with specified return intervals. Facility sizes may then be selected with confidence as to the risk of failure (overflow or backups). If this is done both before and after a rehabilitation program, the frequency distributions or peaks at a specified return period can be used to measure RDII removal effectiveness.

Because the models can provide insight as to the source of extraneous flow, their use can also focus subsequent RDII field investigations in the most cost-effective manner. For example, if the models predict very little inflow, then it is known that sewer televising is in order rather than smoke testing for inflow sources and visa versa. Because the models may be in error as discussed later, such preliminary findings must be used with caution and confirmed in the field.

The continuous simulations are typically run with a 10-minute to one-hour time step, driven by continuous rainfall data and calibrated with measured sewer flows. Models are calibrated to measured sewer flow using rainfall data collected near the study area.

Once calibrated, long-term simulations are performed using a large amount of local meteorological data. Using the model results, flow statistics can be developed such as average annual flows, average dry weather flows (ADWFs), and average WW RDII flows. With two models, one calibrated to pre-rehabilitation data and another calibrated to post-rehabilitation data, the removal effectiveness can be calculated under a variety of flow conditions. Depending on the approach to RDII removal, the RDII flow removal effectiveness may not be consistent for all conditions.

Method 5 determines the variability in RDII removal by performing statistical comparisons for combinations of RDII flow return periods duration. Flow return periods indicate the annual probability of a specified flow event to occur in any given year.

Table 4-2 shows the annual probability of events with different return periods being exceeded in a given year.

Table 4-2. Annual Probabilities of Specific Return Periods.

Return Period	*Annual Likelihood of Exceedence*
1	99%
2	50%
10	10%
25	4%

Flow duration refers to how long the specified flow rate is exceeded. Examples include peak hour, peak day, and peak month. The continuous simulation models perform RDII calculations for each time step that meteorological data are available. Over a long-term simulation, the calculation at each time step is saved for further statistical analysis.

Several methods are available to compute return periods from the database of peak flows. One such method is termed the Log Pearson Type III (LP3) distribution (see *Hydrology Subcommittee of the Intragency Advisory Committee on Water Data, Bulletin 17B*, United States

Geological Survey (USGS) 1982). This statistical distribution is the standard for flood frequency statistical analysis. This choice assumes that the statistical distribution of peak flows in sewers follows that in river systems. This may be in error; however, there have been no studies to define an alternative distribution that is more suitable. The determination of peak RDII removal herein is based on comparison of peak flows with return periods of 5–10 years from 40 years of record or more. At these return periods, the difference in results from available alternative methods of analysis is minor.

4.5.1 Use of LP3 Distributions to Characterize RDII Flows

Predictive I/I models can generate large flow datasets. A proper statistical analysis, however, is necessary to reduce the model output into the information wastewater that planners really must know: "How much flow does the system need to handle for a particular level of service?" Numerous statistical distributions have been suggested on their ability to fit flood data. For decades, federal agencies have used the LP3 distribution to fit the relationship between flow volume and recurrence interval for river systems. As such, the LP3 distribution has become a standard method in engineering.

An LP3 distribution analysis is prepared by following these steps:

1. ***Compute the peak annual series from the model output.*** This simply involves culling the largest flow values from each calendar year (or water year) into a separate data series. This greatly reduces the amount of data handling. The series will be based on whether the user is interested in peak hourly, peak daily, or peak monthly data. Summarizing data into the properly resolved time step must be done first.
2. ***Rank the peak annual flow events and compute the "plotting position" of each event.*** The plotting position or recurrence interval is the average period over which a particular flow would be equaled or exceeded. For example, a 10-year flow would be equaled or exceeded an average of once per 10 years. The Cunnane plotting position formula is shown below:

$$T_R = \frac{\#Years + 0.2}{Rank - 0.4} \qquad (4.5.1\text{-}1)$$

The recurrence interval for a particular event, TR, is roughly equal to the number of years of record divided by the rank of the event. The 0.2 and 0.4 factors in the numerator and denominator, respectively, help correct for the limited size of any sample set. The effects of this correction become less apparent for larger sample sizes or less extreme flow events. For example, the highest ranking event in a 50-year data series would have an estimated recurrence interval of 83.7 years using the Cunnane plotting position, while the tenth largest event would have a recurrence interval of 5.2 years.
3. ***Compute the LP3 fit to the peak annual flow series.*** The LP3 statistical distribution is computed as follows:

$$\log(Flow) = \overline{\log(Flow)} + K\sigma_{\log(Flow)} \qquad (4.5.1\text{-}2)$$

First, calculate the base 10 logarithm of each event in the peak annual series. $\overline{\log(Flow)}$ is the average of the base 10 logarithms of all events in the peak annual series. (Note: The calculation of $\overline{\log(Flow)}$ above is *not* the same as computing the average of the flow values in the peak annual series and then taking the logarithm of this average.) $\sigma_{\log(Flow)}$ is the standard deviation of the set of log (*Flow*) data. The standard deviation can be calculated using a spreadsheet program, or otherwise as:

$$\sigma_{\log(Flow)} = \sqrt{\frac{\sum(\log(Flow) - \overline{\log(Flow)})^2}{\# \ Years - 1}} \qquad (4.5.1\text{-}3)$$

K is the cumulative probability distribution function for the LP3 distribution. It is a complex formula that requires Skew and the *Standard Normal Inverse Probability function. K* values for specific recurrence intervals are typically read from tables in hydrologic texts, such as USGS Bulletin 17B.

4. ***Plot the peak annual flow series and LP3 distribution together.*** The LP3 distribution plot can be particularly useful for smoothing the peak annual data series and predicting the magnitude of infrequent events.

Each case study, except McCandless Township, was analyzed using the LP3 statistical comparison.

Employing the control basin technique described in other methods can further enhance Method 5. With Method 5, an additional continuous model is calibrated to flow data in a basin in which no rehabilitation took place. The model is calibrated to measured data during the pre-rehabilitation period for the rehabilitation basin. The model simulation is then compared to flow and rainfall data collected during the post-rehabilitation period. If the model departs significantly from the measured flows during this second period, it indicates that the models for the rehabilitation basin must be adjusted. Reasons for this occurring are usually related to significantly different rainfall and groundwater conditions between the two monitoring periods. King County, Washington, used a control basin analysis to improve confidence in estimates of RDII removal in the pilot project basins.

4.5.2 Advantages

This method has many advantages. It provides the most complete picture of RDII flow conditions before and after rehabilitation. Flows can be reported with any combination of flow return period-duration combination. The models also provide a time series of flow data that can be used in a dynamic hydraulic model of the collection system.

4.5.3 Disadvantages

The method relies on a significant amount of flow-monitoring data for an accurate analysis of RDII conditions. Ideally, two full wet seasons of data are preferred in each pre- and post-rehabilitation period to provide RDII flow estimations with confidence. Depending on the model used to produce the RDII estimates, meteorological data—in addition to rainfall—is required for the calibration periods. The method also relies on data from a representative long-term meteorological station for performing the long-term simulations. Use of a control basin can mitigate the need for longer flow-monitoring programs.

4.5.4 Limitations

Sophisticated hydrologic models of RDII partition rainfall into direct inflow, evapotranspiration, soil moisture, groundwater, infiltration through sewer cracks and defects, and groundwater export from the basin. The partitioning is for the most part empirical, requiring considerable high-quality flow and rainfall data to calibrate. This means there are several limitations to be considered in their use, as discussed in the following paragraphs.

Because of the several pathways for rainfall to follow, it is possible to achieve a good match between measured and simulated flows for a single isolated event or short monitoring periods with widely varying choices for the partitioning coefficients. Without confirmation through matching model to measurement over multiple events of varying magnitude, the confidence in the model to accurately project results to infrequent events is severely reduced. Confidence in projection of response to infrequent events expands as the number and magnitude of calibration events increase. Where little data is available, particularly with only smaller rainfall events, it is perhaps better to use another analysis method and avoid projection much beyond the range of measured rainfall.

The model results are highly dependent on antecedent rainfall, which drives soil moisture and groundwater buildup. If rainfall patterns in pre- and post-rehabilitation periods are significantly different (a drought occurs in one or the other periods for example), the model for the drought period may not include terms that are dependent on the buildup of soil moisture—making projection to normal rainfall characteristics tenuous at best. As a result, it is important to collect flow data from periods that are reasonably similar. Other checks on the results should also be made. These may include the R-value analysis discussed above, plots of RDII peaks against the summation of antecedent rainfall (Q versus I plots), and plots of peak flows in rehabilitation basins against those in control basins. Although these checks are not substitutes for the modeling approach, they can be used to confirm the modeling results.

Models contain terms representing direct inflow. Response to rainfall in a saturated ground condition, however, can appear to the analyst's eye to be from inflow because of the quickness of the response. There is a danger that the analyst will use the direct inflow terms of the model to fit these events when, in fact, the sources may represent indirect pathways through soil moisture transport. Since direct inflow sources are often relatively inexpensive to correct, some may use these modeling results to the faulty conclusion that high RDII can be cheaply eliminated. While the model results can be used to focus subsequent field investigations, it is vital for suspected inflow sources to be confirmed in the field before program directions are set.

4.6 RDII Reduction Evaluation Methods—Summary

Evaluation of RDII reduction depends on the method used to quantify RDII in the system. Not all quantification methods suit all evaluation methods. For example, if the constant unit rate method was used to quantify RDII in a particular system, Method 1, Sum of RDII Source Correction Efforts Multiplied by Unit Reduction Rates, is the only method that would apply for calculating the RDII removed.

Table 4-3 indicates which evaluation methods are appropriate given the quantification methods.

Table 4-3. Comparison of RDII Modeling and Evaluation Methods.

Quantification Method	*RDII Reduction Evaluation Method*				
	Method 1: Sum of RDII Source Correction Efforts Multiplied by Unit Reduction Rates	*Method 2: Direct Comparison of Pre- and Post-Construction Flow-Monitoring Data*	*Method 3: Comparison of Rehabilitation with Control Basin R-Factors*	*Method 4: R-Factor Correction Analysis with R-Factor Envelope*	*Method 5: Statistical Comparison of Continuous Simulation Models*
Constant unit rate	✓				
Percentage of rainfall volume (R-value)			✓	✓	✓
Percentage of streamflow			✓		✓
Synthetic unit hydrograph			✓	✓	✓
Probabilistic			✓	✓	✓
Predictive equations based on rainfall/flow regression					✓
Predictive equations based on synthetic streamflow and basin characteristics					✓
RDII as component of hydraulic software			✓	✓	✓

Note: A check in the cell indicates the evaluation method can be used with the quantification method.

Method 5, Statistical Comparison of Continuous Simulation Models, is considered the most suitable for determining RDII flow removal success. It provides the best picture of RDII conditions under a variety of flow and rainfall conditions.

CHAPTER 5.0

Case Study Summaries

The case studies presented in this report were selected based on the availability of sufficient information to evaluate the RDII rehabilitation methods employed for each project and to predict the effect of specific sewer rehabilitation techniques on peak and average RDII.

Table 5-1 presents the case studies evaluated in Chapter 5.0, as well as the rehabilitation techniques employed by each.

Table 5-1. Case Studies Evaluated.

Case Study (number of projects under consideration)	***Rehabilitation Technique(s) Employed***
Oak Creek	Basinwide house foundation drainage system construction
Honolulu (2)	Comparison of sewer main rehabilitation with spot repairs and manhole rehabilitation
LOTT (3)	Comparison of sewer main rehabilitation, manhole rehabilitation, and lower house lateral rehabilitation with complete system rehabilitation and upper house lateral rehabilitation only
King County (2)	Analysis of complete basin system replacement: sewer mains, manholes, and full house laterals
McCandless (3)	Sewer main rehabilitation, lower lateral rehabilitation, manhole rehabilitation
Milwaukee	Manhole rehabilitation

For each case study, the following information is discussed:

- Project background
- Monitoring data
- SSES information
- Project details
- Post-construction data analysis
- Conclusions

5.1 Oak Creek Case Study

This case study analysis presents the results of a pilot rehabilitation study in Oak Creek, Wisconsin, that MMSD and the city of Oak Creek conducted. Rehabilitation efforts in this basin consisted of constructing a stormwater interceptor in a 51-acre residential neighborhood to which foundation drains from adjacent properties were directed. City records do not indicate which of these foundation drains was connected to the sanitary sewer before the project. Given the age of the homes in the basin, any connections that were present would have been illegal, as state plumbing code would not have allowed the practice. The importance of this project analysis is that it provides insight on the amount of RDII remaining after all possible defects in the basin had been addressed, except the house lateral itself.

5.1.1 Project Background

Oak Creek, Wisconsin, is in the greater Milwaukee metropolitan area and receives regional wastewater services from MMSD. The project basin, which Figure 5-1 shows, is situated in the Chapel Hills subdivision in northern Oak Creek. An 8-inch diameter sanitary sewer, with a total length of just less than 7,100′, serves the 51-acre fully residential basin. According to the city of Oak Creek, the basin was one of the higher RDII basins in the city, with a reported peak-daily (24-hour total) to base dry weather flow ratio of 10.8:1, as reported in the MMSD 2010 Facilities Plan. Construction of the 118 homes in the basin occurred in two phases, with the majority of homes constructed between 1966 and 1970. A second phase, comprising a single street, was developed in 1987. The Wisconsin State Plumbing Code was changed in 1955 to make illegal any foundation drain and sump pump connections to the sanitary sewer in new construction. The topography of the land is flat, with well-drained to somewhat poorly drained soils that have a subsoil of silty clay loam and silty clay (*Soil Survey of Waukesha and Milwaukee Counties Wisconsin*, United States Department of Agriculture (USDA), 1998).

Table 5-2 summarizes the characteristics of the Oak Creek study basin.

Table 5-2. Characteristics of Oak Creek Study Basin.

Basin	*Area (acres)*	*Population*	*Sewer System (feet)*	*House Laterals*	*Sanitary Sewers (idm)*	*Number of Manholes*
			8-Inch Diameter			
Chapel Hills	51	Unknown	7,100	118	10.76	Unknown

Figure 5-1 shows the Oak Creek study area.

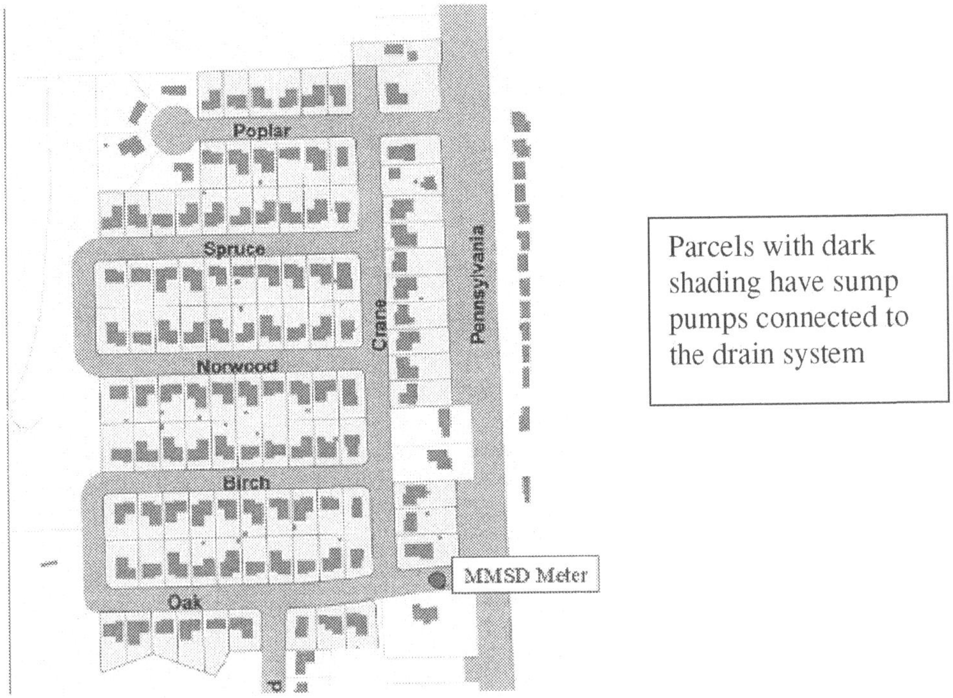

Figure 5-1. Oak Creek Study Area.

5.1.2 Monitoring Data

This section describes the flow and rainfall data collected as part of this sewer rehabilitation project.

5.1.2.1 Flow Data

Flow-monitoring data was collected both before and after rehabilitation. Data before the rehabilitation was collected from July 1995 through July 1996; data after the rehabilitation was collected from July 20, 2000, through September 18, 2000.

Before construction, a depth-monitoring device was installed in the pipe at the downstream end of the basin to measure flows during dry weather and WW. Flows were then computed using Manning's equation and pipe characteristics. The practice of measuring only depth to determine flow can be subject to error if a backwater hydraulic condition prevents the pipe from flowing at normal depth. Review of existing information and discussions with MMSD monitoring staff suggest that such conditions did not occur during the monitoring period. The basin, which is a short distance downstream of the monitoring location, discharges flows to a deep interceptor. MMSD staff discussions indicate that although this interceptor is known to surcharge under extreme flow conditions, it is incapable of surcharge to a level that would induce backwater in the project basin. For the pre-rehabilitation period, the monitoring record ranges from July 21, 1995, to July 17, 1996. Although initial data collections were taken at a more frequent time step, the MMSD database stored the recorded average depths and estimated flows at a one-hour time step.

After construction, an Isco 4150 area-velocity flow meter monitored the flows for the project site. Velocity, wastewater levels, and calculated flows were gathered during the monitoring process. On a weekly basis, MMSD field crews took manual depth measurements and compared them with monitor measurements. When crews determined there was a large enough difference between the two, the meter depth was reset to the manual measurement. Crews also cleared the velocity and depth sensors of any apparent debris during each visit. Data was collected on a five-minute time step.

5.1.2.2 Rainfall Data

Rainfall information was acquired from NOAA for Mitchell Field (the local airport) from 1940 through 2001. Mitchell Field is approximately 2.6 miles from the project site and, over the long term, experiences similar weather patterns to those experienced in the project basin. Mitchell Field served as both the calibration and long-term rain gauge.

5.1.3 SSES Information

In 1999, limited SSES (LSSES) work was performed, which included flow monitoring and manhole inspections. The LSSES identified several problem manholes in the basin.

5.1.4 Project Details

This section details the work performed during this sewer rehabilitation project, including a discussion of project costs.

5.1.4.1 Construction Details

In 1996, the city of Oak Creek completed an RDII removal project that installed 7,273 LF of 6″ PVC storm drain for the connection of foundation drains from residential lots. The 6″ drain was placed in the public right-of-way (PROW) line on both sides of the street, on the house side of the sidewalk. A 4″ capped connection was provided for each lot. Capped cleanouts were provided at intervals of 90–200′, with most not more than 150′ apart. The length of each section of the public drain system ranged from 17 LF (service for a single lot) to 570 LF. Most were connected to existing storm inlets at the curb line, which drained to the existing storm sewer system. Three reaches were directed to existing ditch outlets.

The drain lines were placed with the pipe invert approximately 2′ below grade, following the contour of the ground surface. The pipe was placed by open cut method, except at concrete driveway and sidewalk crossings, which were augured. The open cut pipe, with a total length of 5,320 LF, was backfilled with crushed stone and covered with 6″ of topsoil and sod. Augured pipe length was 1,538 LF. An additional 410 LF of pipe connecting drain reaches to ditch outlets were backfilled with native soil and topsoil and then seeded. Drain lines in the subject study area were installed under three different construction contracts. Subsequent to public construction, private property owners were required to install connections from their foundation drains to the 6″ drain system. The city did not provide funding assistance to property owners to complete the drainage connections. City records indicate that of the 118 homes in the basin, approximately 112 were connected to the drain system.

The portion of the work in the PROW was completed between 1996 and 1998 as part of three different construction projects. Information provided by the city was in the form of construction plans and specifications, bid tabs, and payment requests. Record drawings were used where available to confirm actual material quantities installed. The city also provided some additional summary information on request. The city engineering department performed the design, contract administration, and inspection.

5.1.4.2 Construction Costs

Projects were awarded to low bidders based on unit cost totals for the estimated quantities shown on the construction documents. The contractors' unit prices included all general and incidental costs.

Table 5-3 summarizes the total costs for the project.

Table 5-3. Oak Creek Construction Cost Summary.

Item	*Description*	*Quantity*[a]	*Unit*	*Unit Cost*[b]	*Total*[c]
1	6-inch PVC storm sewer, crushed stone backfill, 6-inch topsoil and sod restoration	5,323	LF	$20.15	$101,857
2	6-inch PVC storm sewer, augured under driveways and sidewalks	1,538	LF	$44.35	$57,640
3	6-inch PVC storm sewer, ground backfill, Type C surface restoration	413	LF	$18.50	$7,313
4	Storm sewer connection to existing inlet	23	Ea	$500.00	$9,140
5	7-inch concrete pavement removal and replacement	5.44	SY	$80.00	$435
6	Connection of existing sump drains to new system	15	Ea	$20.00	$300
Total Construction Cost					$176,685
Total Construction Cost Adjusted to 2002 Dollars[d]					$199,556

[a]Sum of quantities for all three phases of work.
[b]Represents average bid unit cost for all three phases. Costs reflect raw dollars, unadjusted for inflation.
[c]Represents total cost for all three phases. Costs reflect raw dollars, unadjusted for inflation.
[d]Projects were constructed over three consecutive years and adjusted according to 20-city average *Engineering News-Record* (ENR) construction cost index (ENR 5825). Does not include an estimated $38,000 in homeowner costs to connect to the system.

The costs presented in Table 5-3 do not include the cost of each homeowner's connecting the foundation drain to the system. The city has estimated that each homeowner spent an average of $300 per connection. Engineering and administrative costs were approximately 8% of the construction contract.

5.1.4.3 Review of Costs

The project features were limited in scope. The new drainpipe system was designed specifically to provide a path for foundation drainage to the storm sewer system, which was otherwise being directed into the sanitary sewer system. The total public cost in 2002 dollars was about $4,000 per acre. Having the project designed and administered by city personnel can lead to cost savings for municipalities that have this capability.

Locating the drain within the PROW, but mainly in unpaved areas, reduced the need for traffic control and greatly simplified surface restoration. Placing the pipe under driveways and sidewalks by auguring instead of open cut took advantage of these same factors. The shallow depth of the excavation also meant that there was little disturbance to existing utility lines.

Finally, the area was already served by a separate storm sewer system. This new drain system was an enhancement beyond simply placing the sump pump discharge outside the house in the hope that the flow will eventually find the nearest catch basin or infiltrate into the soil.

5.1.5 Post-Construction Data Evaluation

This section discusses the analysis performed before and after construction conditions. The discussion includes a review of data quality issues and an overall summary of RDII reduction success.

5.1.5.1 Data Quality

An examination of the flow and rainfall data indicates several concerns that may affect the results of the analysis:

1. Flow data from the pre-rehabilitation period was collected from a depth-only monitor (submerged pressure sensor). Flows were computed using Manning's equation. There may be some errors in the results due to the lack of corresponding velocity data; however, anecdotal evidence suggests that this concern may not be warranted because of downstream hydraulic conditions.
2. Flow response to rainfall was not always consistent. For example, Figure A-1 in Appendix A shows that the peak flows on August 19 and August 30, 1995, are significantly higher than immediately preceding events with greater rainfall. Because the peaks and recession limbs of the preceding events are well matched by the RDII model, the peaks on the 19th and 30th are not believed to reflect the rainfall data used in the analysis. This may be due to two factors, rainfall errors being the most probable:
    a. Rainfall measured did not accurately reflect what actually fell on the pilot basin.
    b. Downstream hydraulic conditions affected the depth of flow, which in turn affected the estimated flows from the basin.
3. Note that rainfall of about 1.3″/hr (highest measured) resulted in similar flow peaks on August 16, 1995, and June 18, 1996 (Figures A-1 and A-2, respectively). The model accurately represents these events.
4. In the post-rehabilitation data, there are notable periods where average flows decrease sharply from one day to the next. This may indicate debris hanging up on the velocity sensors, which may affect subsequent peak flows. Observation of the subsequent recessions, however, suggests that the debris was cleared during the higher flows.

5.1.5.2 Calibration

Graphs comparing the model predictions with measured flow and rainfall appear as Figures A-1 through A-4 in Appendix A. In each monitoring period, significant peaks occur that could be used for calibration (e.g., August 30, 1995, and August 5, 2000). Calibrating the model against these peaks, however, results in the model's overpredicting flows for the other smaller peaks with an error of 50–100 percent. As discussed earlier, the peak on August 30, 1995, is in question. In addition, the peak on August 5, 2000 (post-rehabilitation period), exceeds the peaks in the pre-rehabilitation period even though the 2000 peak was in response to less rainfall (0.9 versus 1.3″/hr).

Because of concerns raised earlier, the model calibration focused on the most consistent peak flows. Calibration of the pre-rehabilitation model focused on peaks that occurred on August 16, 1995, and June 17, 1996. The post-rehabilitation model focused on achieving reasonable matches to flows measured on August 17, 2000, and September 12, 2000. In the post-rehabilitation period, this approach resulted in the model's underpredicting the peak on August 5, 2000, by about 30%. Table A-1 in Appendix A presents the errors in predicted versus measured flows for the periods surrounding rain-induced peaks. The overall error in volume for the complete periods of analysis averaged about 20%.

5.1.5.3 Results

During the calibration periods, the model suggests that RDII measured during the monitoring periods dropped from 16% of total measured flow volume to 9% after the rehabilitation. In addition, the percentage of rainfall capture to sewer flow dropped from 16% to 1%. Because the rainfall was significantly different in the two periods of analysis, these figures are useful only as a guide. Long-term rainfall series results are discussed later.

To define the differences between the periods of analysis, the models were used to compute hourly RDII for each hour of the 57-year rainfall record at the Mitchell gauge. The annual maximum peak hourly RDII, as well as the maximum annual calendar day RDII (24-hour average), were extracted from the resulting databases and used to prepare recurrence relationships for each model. The recurrence relationships were developed by fitting an LP3 curve to the annual maximum data. In this way, it is possible to compare peak RDII with defined return periods for both the pre- and post-rehabilitation periods, providing an accurate estimate of the reduction in RDII and peak RDII.

Tables 5-4, 5-5, and 5-6 present the percentage reductions in peak RDII for defined return periods (i.e., peak hour, peak day, and peak month, respectively). A review of the measured peak RDII flows compared to the antecedent rainfall (see Figure A-6 in Appendix A) supports the removal estimate.

Table 5-4. Pre- and Post-Construction Peak-Hour RDII Conditions for Oak Creek Basin.

Condition	*5-Year Peak-Hour RDII*			*10-Year Peak-Hour RDII*			*20-Year Peak-Hour RDII*		
	mgd	*gallons per acre per day (gpad)*	*gpd/idm*	*mgd*	*gpad*	*gpd/idm*	*mgd*	*gpad*	*gpd/idm*
Pre-Construction	0.279	5,474	25,953	0.321	6,285	29,798	0.359	7,046	33,402
Post-Construction	0.145	2,851	13,517	0.165	3,244	15,379	0.185	3,624	17,182
RDII Removal Effectiveness	0.134	2,623	12,436	0.155	3,041	14,418	0.174	3,421	16,221
Percent Removal Effectiveness	48%			48%			49%		

Table 5-5. Pre- and Post-Construction Peak-Day RDII Conditions for Oak Creek Basin.

Condition	*5-Year Peak-Day RDII*			*10-Year Peak-Day RDII*			*20-Year Peak-Day RDII*		
	mgd	*gpad*	*gpd/idm*	*mgd*	*gpad*	*gpd/idm*	*mgd*	*gpad*	*gpd/idm*
Pre-Construction	0.195	3,814	18,083	0.226	4,423	20,967	0.255	5,005	23,730
Post-Construction	0.060	1,178	5,587	0.067	1,318	6,248	0.074	1,445	6,849
RDII Removal Effectiveness	0.134	2,636	12,496	0.158	3,105	14,719	0.182	3,561	16,881
Percent Removal Effectiveness	69%			70%			71%		

Table 5-6. Pre- and Post-Construction Peak-Month RDII Conditions for Oak Creek Basin.

Condition	*5-Year Peak-Month RDII*			*10-Year Peak-Month RDII*			*20-Year Peak-Month RDII*		
	mgd	*gpad*	*gpd/idm*	*mgd*	*gpad*	*gpd/idm*	*mgd*	*gpad*	*gpd/idm*
Pre-Construction	0.065	1,280	6,066	0.069	1,346	6,384	0.072	1,403	6,650
Post-Construction	0.027	530	2,512	0.027	537	2,548	0.028	544	2,578
RDII Removal Effectiveness	0.038	750	3,555	0.041	809	3,836	0.044	859	4,072
Percent Removal Effectiveness	59%			60%			61%		

Figures 5-2, 5-3, and 5-4 show the results for peak hourly, peak daily, and peak monthly RDII, respectively.

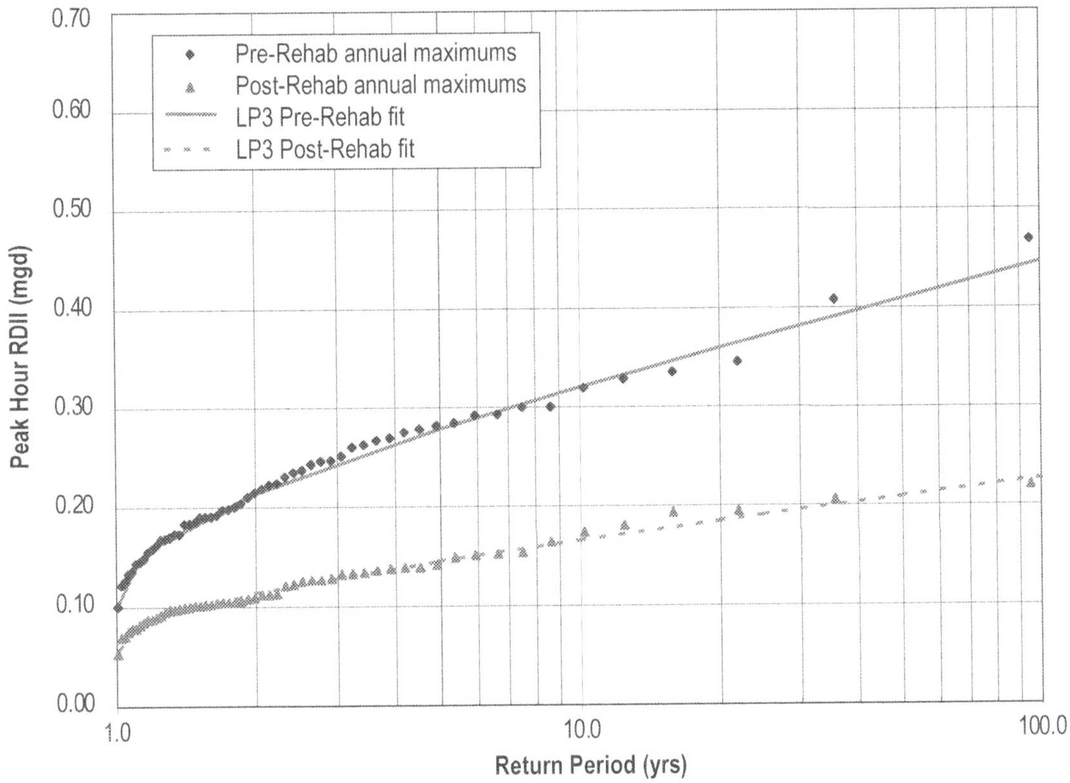

Figure 5-2. Return Period for Peak-Hour RDII for Oak Creek.

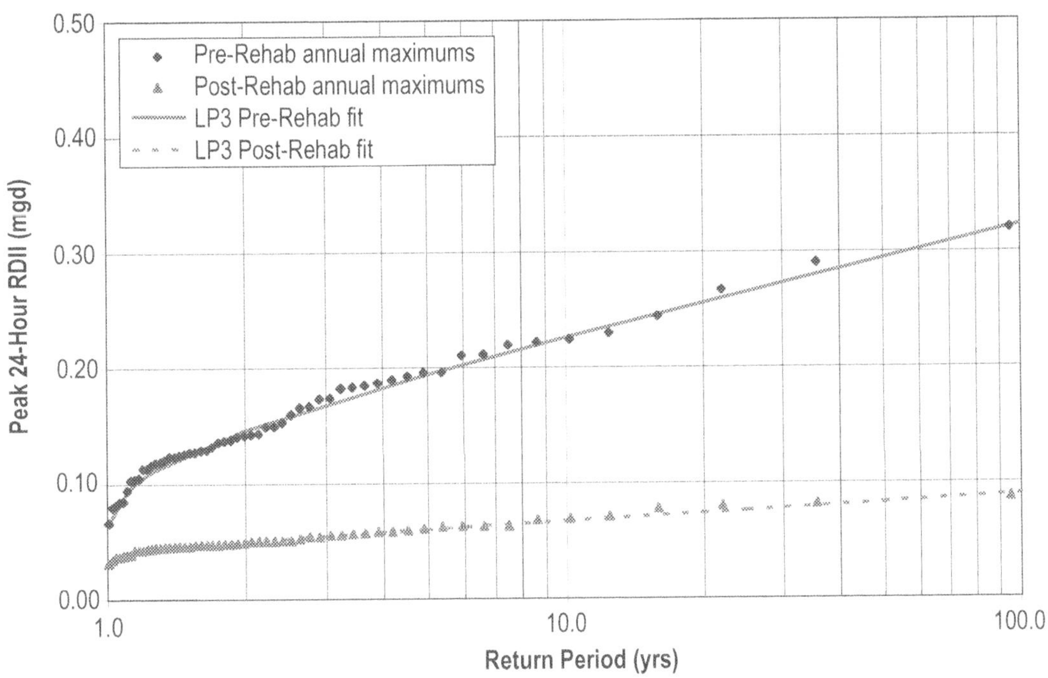

Figure 5-3. Return Period for Peak-Day (24-Hour Average) RDII for Oak Creek.

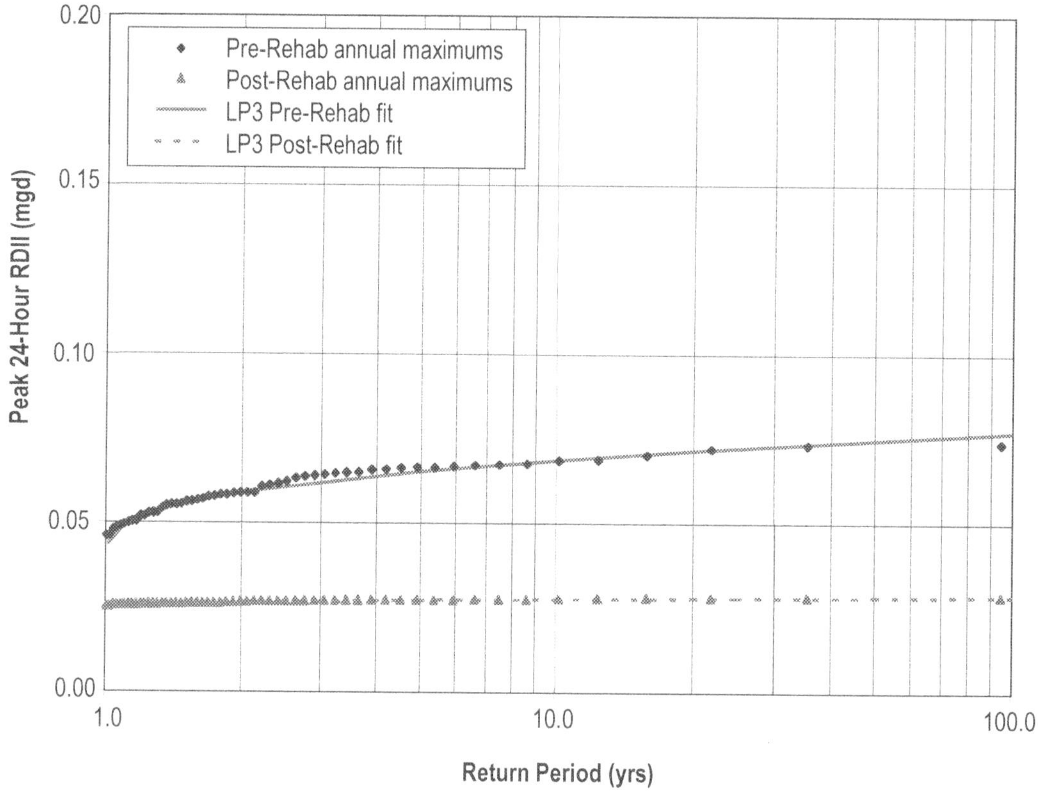

Figure 5-4. Return Period for Peak-Month (30-Day Average) RDII for Oak Creek.

5.1.5.4 Cost-Effectiveness Analysis

The cost-effectiveness analysis of this project combines the cost data gathered with the estimated RDII reductions achieved.

Table 5-7 summarizes the estimated cost-effectiveness for the Oak Creek project in terms of peak RDII frequency and duration.

Table 5-7. Peak RDII Removal Cost-Effectiveness Measures for Oak Creek.

RDII Condition	RDII Removal Cost, $/gpd [a,b]		
	Peak Hour	Peak Day	Peak Month
5-year	$1.49	$1.48	$5.22
10-year	$1.29	$1.26	$4.84
20-year	$1.14	$1.10	$4.56

[a]Construction costs only.
[b]Costs are adjusted to 2002 dollars.

Determining the cost-benefit of the Oak Creek project requires three values: treatment cost avoided, storage cost avoided, and conveyance cost avoided. For a proper comparison, the peak-day RDII reduction should be used for treatment and storage cost-avoidance comparisons, and the peak-hour value should be used for the conveyance cost avoidance.

According to the 2001 Proposed O&M Budget, MMSD O&M treatment costs for average RDII are approximately $500 per million gallons of wastewater. In this light, the avoided treatment costs do not appear to warrant the expenditure.

The MMSD 2010 Facilities Plan estimated, however, that it would cost approximately $177.5 million (not adjusted) to provide an additional 420 acre-feet of tunnel storage that would reduce the overflow frequency from once in five years to zero in five years. This storage cost translates to $9.70 per gallon stored (unadjusted). This value can be compared directly with the peak-day RDII reduction in the five-year event for the Oak Creek project at $1.48 per gallon of peak-day flow.

A third comparison can be made with conveyance costs. The MMSD 2010 Facilities Plan also estimated the Oak Creek south interceptor would cost approximately $4.05 million to provide 10 cubic feet per second (CFS) capacity to south central Oak Creek. The cost per RDII rate for the facilities plan converts to $0.63 gpd peak RDII rate, compared with a peak-hour RDII rate reduction of $1.49 for the five-year event from the Oak Creek project.

5.1.6 Conclusions

Overall, it appears that the project successfully removed RDII to a point that reduced cost-effective peak RDII flows, in terms of both hourly rates and daily averages. The project appears to have worked well because it makes groundwater and rapidly infiltrated rainfall unavailable for migration into pipe defects present in the sewer lateral.

5.2 Honolulu Case Study

The purpose of this case study analysis is to summarize the effectiveness of the pilot project resulting from the Sewer Rehabilitation and I/I Minimization Program (RDII Program) for the city and county of Honolulu. The pilot project (September 1995) was planned for comprehensive repairs to be completed in one basin (Basin A) and spot repairs to be completed in a second nearby basin (Basin B). Both basins were considered to be WW SSO basins.

5.2.1 Project Background

Figure 5-5 shows the Honolulu study area.

Basin A. Basin A is a sanitary sewer basin located in the low-lying area of the Enchanted Lake area of Kailua, Oahu, in the Hawaiian Islands. The Enchanted Lake area was selected for rehabilitation because the area is known to have frequent rainfall. Flow monitoring had shown evidence of high RDII. Basin A comprises single-family residences on 203 parcels, totaling 49 acres. The sewer mains are 8″ diameter vitrified clay pipe (VCP), totaling 8,333 LF. Manholes number 55, 28% of the sewer mains are below sea level, and laterals are 6″ diameter VCP. The sewer lines were constructed between 1964 and 1971. In 1995, the average age of the collection system was 27 years old.

Before the RDII Program, main line joints were grouted in Basin A during the Ahuimanu/Kaneohe/Kailua sewer rehabilitation project (1992). There were 238 joints grouted out of a total of 1,630 joints, equaling 15% of the joints. Grouting was used to seal defective joints on the

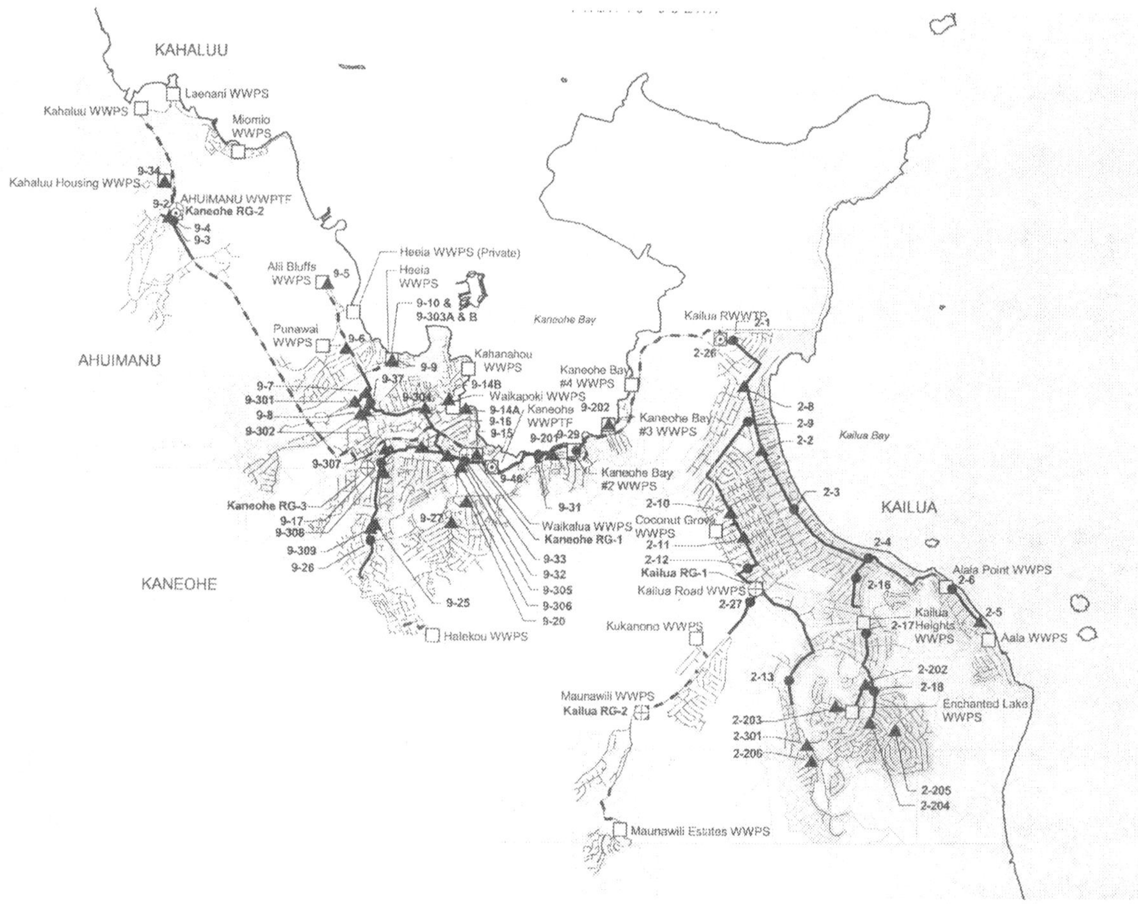

Figure 5-5. Honolulu Study Area.

main sewer line. In addition, manhole inserts were installed in all manholes to prevent inflow from entering the collection system through the manhole lids.

As part of the RDII Program, the city allowed design/build teams to investigate and submit various rehabilitation methods. The city specified, however, that only long-term rehabilitation methods (i.e., 50 years or more) would be considered acceptable for the pilot project. Short-term repair measures such as grouting were not considered acceptable.

The city determined that a variety of long-term methods was acceptable, including CIPP, fold-and-form pipe, and pipe replacement by pipe bursting or traditional open cut methods.

For Basin A, a comprehensive approach comprising the rehabilitation of all main lines and laterals within a rehabilitated basin was planned. In February 1995, a request for proposal (RFP) was advertised for the pilot project. The RFP required a full-service contractor to furnish all necessary labor, equipment, materials, tools, supplies, and accessories to design and construct rehabilitation improvements to the existing sewer system.

On April 21, 1995, a contract was awarded to use the CIPP for the main lines and laterals in both pilot project basins. After contractor selection, additional city requirements were developed that disallowed trenching or access pits for the lateral rehabilitation. Because of these requirements, lateral rehabilitation was subsequently deleted from the pilot project.

Basin B. Basin B is a sanitary sewer basin located in the upper-sloped area of the Enchanted Lake area of Kailua, Oahu, in the Hawaiian Islands. The Enchanted Lake area was selected because it is known to have frequent rainfall, and flow monitors in a residential area had recorded high levels of RDII. Basin B comprises single-family residences on 156 parcels, totaling approximately 36 acres. Sewer mains are 8″ diameter VCP, with a total length of 5,425 LF. Manholes number 38 and the average lateral length per parcel is 24 LF. The lower laterals—those in the PROW that connect to the sewer mains—are 6″ diameter VCP. The sewer lines were constructed as part of four development projects, which were completed between 1958 and 1984. In 1995, at the outset of the pilot project, the average age of the collection system in the project area was 28 years.

Before the RDII Program, main line joint grouting was completed in Basin B in 1992 and 1993 as part of the Ahuimanu/Kaneohe/Kailua sewer rehabilitation project. Manhole inserts were installed in all manholes. According to Test and Seal Reports, only 2% of the joints in Basin B failed the pressure test and were subsequently grouted. A small portion of the main lines (12 percent by length) was classified as being in fair or fair-to-poor condition based on closed-circuit television (CCTV) inspection. Manhole inspections indicated that the manholes had no structural defects.

To meet a consent decree objective to evaluate different rehabilitation methods, the city allowed design/build teams to investigate and submit various rehabilitation methods. The city specified, however, that only long-term rehabilitation methods (i.e., providing 50 years or more of useful life) would be considered acceptable for the pilot project. Short-term repair measures such as grouting were not considered acceptable.

The city determined that a variety of long-term methods was acceptable, including CIPP, fold-and-form pipe, and pipe replacement by pipe bursting or traditional open cut methods.

The Basin B approach to sewer rehabilitation in the pilot project was spot repairs, which included rehabilitation of only those main lines and lower laterals (within the PROW) known to be defective within a rehabilitation basin. Basin B work required an assessment of the sewer system, including mains, manholes, and lower laterals in the PROW and easement areas. In February 1995, an RFP was advertised for the pilot project. The RFP required a full-service contractor to furnish all necessary labor, equipment, materials, tools, supplies, accessories, designs, and improvements to the existing sewer system.

On April 21, 1995, a contract was awarded to use the CIPP for the main lines and laterals in both pilot project basins. After contractor selection, additional city requirements were developed that disallowed trenching or access pits for the lateral rehabilitation. Because of these requirements, lateral rehabilitation was subsequently deleted from the pilot project.

Table 5-8 summarizes the characteristics of the Basin A and Basin B study basins.

Table 5-8. Characteristics of Honolulu Study Basins.

Basin	Area (acres)	Population	Sewer System (feet) 8-Inch Diameter	House Laterals	Sanitary Sewers (idm)	Number of Manholes
Basin A	49	—	8,333	203	12.63	55
Basin B	36	—	5,425	156	8.22	38

5.2.2 Monitoring Data

This section describes the flow and rainfall data collected as part of this sewer rehabilitation project.

Several long-term flow monitors (the ADS Company) and two rain gauges were installed in Kailua in 1993. Special control basin intensive flow monitors were also installed in the Enchanted Lake area in 1994 for the pilot project. These intensive flow monitors remained in place before rehabilitation, during most of the construction, and subsequent to the construction. The monitoring network included 1) basin monitors 2-203 (Basin A) and 2-205 (Basin B) and 2) intensive monitors 2-204, 2-206 (Basin B), and 2-301. This flow-monitoring network, along with the nearby rain gauge, provided the necessary data for the analysis.

Both pilot basin flow monitors recorded dry weather flows and WWFs with few monitor failures. Some control monitors, however, experienced failures during WW conditions or had no response to rainfall. Monitor failures or unusual data observations were noted following data collection. Rainfall patterns varied greatly before and after construction; however, no rain gauge failures were recorded during the analysis period.

5.2.2.1 Flow Data

The Enchanted Lake area was favored for selection because of the evidence of high RDII as shown in the flow-monitoring data, available from 1992–1994. Before the pilot project, flow monitoring indicated high volumes of RDII. The average minimum flow was measured at 0.03 mgd during the early morning hours. The highest peak flow was measured at 0.57 mgd, which is eight times the average daily flow of 0.04 mgd. This peak flow measurement was taken during heavy rainfall in March 1994.

In Basin B, the average minimum flow was approximately 0.01 mgd, which occurred in the early morning hours. As of 1995, the highest measured peak flow for flow monitor 2-205 was 0.54 mgd, which is about 13 times the average daily flow of 0.04 mgd. This peak flow occurred during heavy rainfall in February 1994.

Basin A. Basin A is located upstream of flow monitor 2-203, which provided pre- and post-construction flow data for Basin A.

Basin B. Basin B is defined by the tributary area upstream of flow monitor 2-205, which provided pre- and post-construction flow data for Basin B.

Significant RDII responses were observed at both basin flow monitors before and after construction. No significant changes in diurnal or minimum flows were observed in the flow-monitoring network; however, RDII responses were highly variable between the basin monitors and control monitors. Monitor 2-203 was temporarily removed to accommodate construction. Data was collected for more than one year following rehabilitation to provide adequate data for analysis.

5.2.2.2 Rainfall Data

Basin A. Basin A is located near Kailua RG-1. This rain gauge was installed in January 1993 and provided rainfall data for the rehabilitation effectiveness analysis.

Basin B. Kailua RG-1 also provided rainfall data for Basin B.

5.2.3 SSES Information

To progress toward upcoming rehabilitation projects in these areas, SSES source-detection field work—including CCTV, visual manhole inspections, smoke testing, and mini-basin flow monitoring—was performed in these basins.

As part of the grouting process, CCTV inspections and visual manhole inspections were performed. The CCTV inspection tapes and reports showed cracked mains that were not addressed by the grouting process. Fifty-nine percent of the sewers' main line length was classified as being in fair or fair-to-poor condition. Manhole inspections during this process indicated that the manhole had no structural defects.

Basin A. In early 1993, smoke testing was conducted in Basin A and Basin B and reported in the *Kailua and Kaneohe-Ahuimanu Smoke Testing Report* by Barrett Consulting Group. The smoke test results indicated that 11% of the parcels had cleanout covers missing or showed evidence of cracked laterals in Basin A.

Basin B. For Basin B, smoke testing revealed that 12% of the parcels had evidence of cracked laterals, missing cleanout covers, or yard drains connected to the sewer system. There were a total of 18 defects out of 156 parcels detected by smoke testing; those defects consisted of five-yard drain connections, 10 cleanout (missing covers), and three cracked laterals.

Other information about the condition of the sewers was obtained through conversations with personnel from the city's Department of Wastewater Management, Collection System Maintenance.

5.2.4 Project Details

This section provides details on the work performed during this sewer rehabilitation project, including a discussion of project costs.

Flow and rainfall data was analyzed before and after rehabilitation construction to determine the initial RDII characteristics and the extent of RDII reduction in the rehabilitated basin. The construction of the pilot project was performed using the innovative design/build process. This approach was selected because of the short time period in which both the design and construction were to be completed. Although new to the city and public works projects, this approach has been widely and successfully used by private industry.

The pilot project area was selected and specifications were developed for acceptable rehabilitation methods. This information was presented in the bid documents, which required the design/construction teams, comprising engineers and contractors, to determine the most appropriate rehabilitation methods for the project.

5.2.4.1 Construction Details

The intent of pipeline and manhole rehabilitation was to repair structural defects and reduce the amount of RDII. It was recommended to rehabilitate all mains, manholes, and laterals except in a significant main line sag area where replacement with a 10″ sewer line was necessary. Laterals along the replacement pipe section were also replaced because of sagging or reverse slope. All manholes located on the sewer main were rehabilitated, using applied coating systems by Sauereisen or Xypex.

Basin A. Rehabilitation of the sewer mains in Basin A was accomplished by installation of CIPP by InLiner. The portion of sewer mains that was repaired consisted of 6,768 LF of CIPP. Lengths of pipe considered for replacement were 865 LF of 8″ diameter and 700 LF of 10″ diameter pipes. The percent of pipe in the basin rehabilitated by length is 81%, and the percent of pipe in the basin replaced by length is 19%. The number of manholes rehabilitated was 44 by Sauereisen, eight by Xypex and Plug, and three by infishhield. The percentage of manholes rehabilitated in Basin A was 95%, and the percentage of manholes replaced was 5%.

Basin B. Basin B rehabilitation involved spot repair of selected sewer mains and manholes. Following detailed CCTV inspection of pipes in Basin B, significant defects were identified for rehabilitation. Significant defects consisted of pipe cracks, offset joints, and extensive root intrusion. Typical rehabilitation consisted of using CIPP by KM Partliner or KM Sideliner at the damaged sections of pipe only. Some pipe reaches had multiple damaged sections and were entirely rehabilitated with CIPP by InLiner. The determination was made not to repair the sag location identified during the project and to ignore minor root penetrations. Only some manholes located adjacent to repaired pipe sections were rehabilitated by plugging leaks and coating with Xypex. No manholes or pipes were replaced.

5.2.4.2 Construction Costs

Table 5-9 summarizes the cost breakdown for each basin.

Table 5-9. Honolulu Construction Cost Summary.

Construction Costs	Basin A	Basin B
Sewer Main Construction Cost		
• Rehabilitation	$663,411	$395,269
• Replacement	$1,032,058	—
Manhole Construction Cost		
• Rehabilitation	$105,149	$13,212
• Replacement	$65,107	—
Total Cost	$1,865,725	$408,481
Total Construction Cost Adjusted to 2002 Dollars	$2,184,094[a]	$515,816[b]

[a]Costs were adjusted according to the 20-city average ENR construction cost index (ENR 5620).
[b]Costs were adjusted according to the 20-city average ENR construction cost index (ENR 5210).

5.2.5 Post-Construction Data Evaluation

This section discusses the analysis performed before and after construction conditions. The discussion includes a review of data quality issues and an overall summary of RDII reduction success.

As stated earlier, Basin A and Basin B appeared to have significant RDII response after rehabilitation. RDII rates and volumes varied, depending on the characteristics of the storm and other factors (i.e., soil conditions, sewer structural condition).

5.2.5.1 Calibration

See Appendix B for calibration plots and other supplementary information on Basin A and Basin B.

5.2.5.2 Results

Basin A. Tables 5-10, 5-11, and 5-12 present the percentage reductions in peak RDII for defined return periods (i.e., peak hour, peak day, and peak month, respectively). Only minor reductions in peak RDII were observed (5% for peak hourly flows). This finding is supported by a review of peak measured RDII compared to antecedent rainfall (see Figure B-19 in Appendix B).

Table 5-10. Pre- and Post-Construction Peak-Hour RDII Conditions for Honolulu Basin A.

Condition	5-Year Peak-Hour RDII			10-Year Peak-Hour RDII			20-Year Peak-Hour RDII		
	mgd	gpad	gpd/idm	mgd	gpad	gpd/idm	mgd	gpad	gpd/idm
Pre-Construction	0.395	8,063	31,291	0.432	8,819	34,225	0.447	9,115	35,376
Post-Construction	0.376	7,666	29,752	0.409	8,354	32,421	0.422	8,617	33,441
RDII Removal Effectiveness	0.019	397	1,539	0.023	465	1,804	0.024	499	1,935
Percent Removal Effectiveness	5%			5%			5%		

Table 5-11. Pre- and Post-Construction Peak-Day RDII Conditions for Honolulu Basin A.

Condition	5-Year Peak-Day RDII			10-Year Peak-Day RDII			20-Year Peak-Day RDII		
	mgd	gpad	gpd/idm	mgd	gpad	gpd/idm	mgd	gpad	gpd/idm
Pre-Construction	0.218	4,454	17,285	0.219	4,466	17,334	0.224	4,563	17,711
Post-Construction	0.211	4,307	16,714	0.212	4,319	16,761	0.217	4,425	17,172
RDII Removal Effectiveness	0.007	147	571	0.007	148	574	0.007	139	539
Percent Removal Effectiveness	3%			3%			3%		

Table 5-12. Pre- and Post-Construction Peak-Month RDII for Honolulu Basin A.

Condition	5-Year Peak-Month RDII			10-Year Peak-Month RDII			20-Year Peak-Month RDII		
	mgd	gpad	gpd/idm	mgd	gpad	gpd/idm	mgd	gpad	gpd/idm
Pre-Construction	0.024	487	1,890	0.025	502	1,947	0.025	502	1,950
Post-Construction	0.023	479	1,857	0.024	493	1,912	0.024	493	1,915
RDII Removal Effectiveness	0.000	9	33	0.000	9	35	0.000	9	35
Percent Removal Effectiveness	2%			2%			2%		

Figures 5-6, 5-7, and 5-8 show the LP3 curves for peak hourly, peak daily, and peak monthly RDII, respectively.

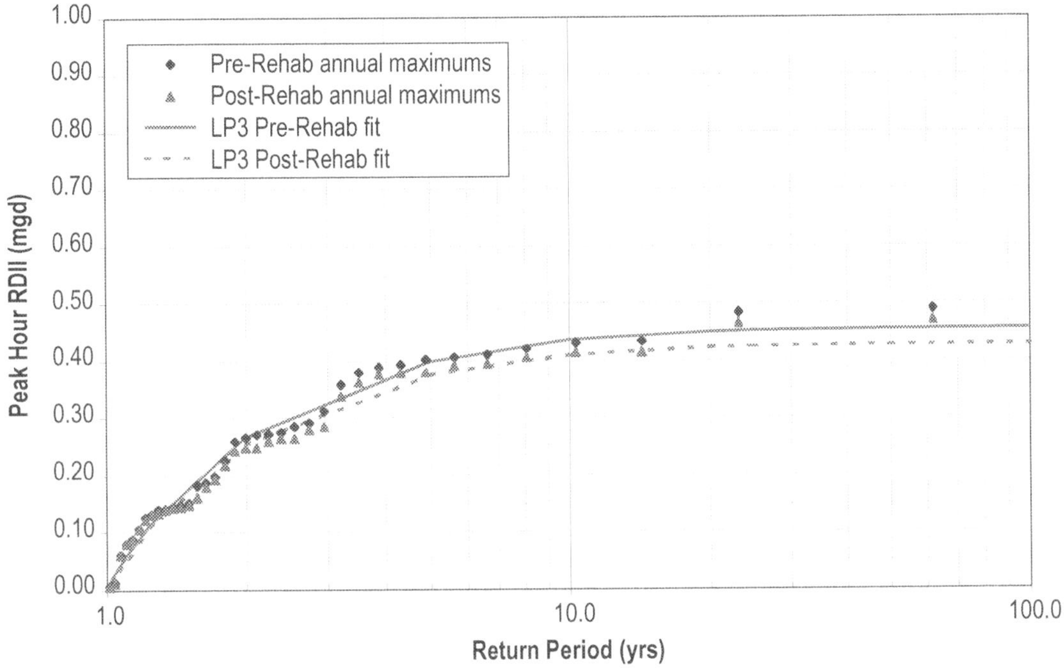

Figure 5-6. Return Period for Peak-Hour RDII for Honolulu Basin A.

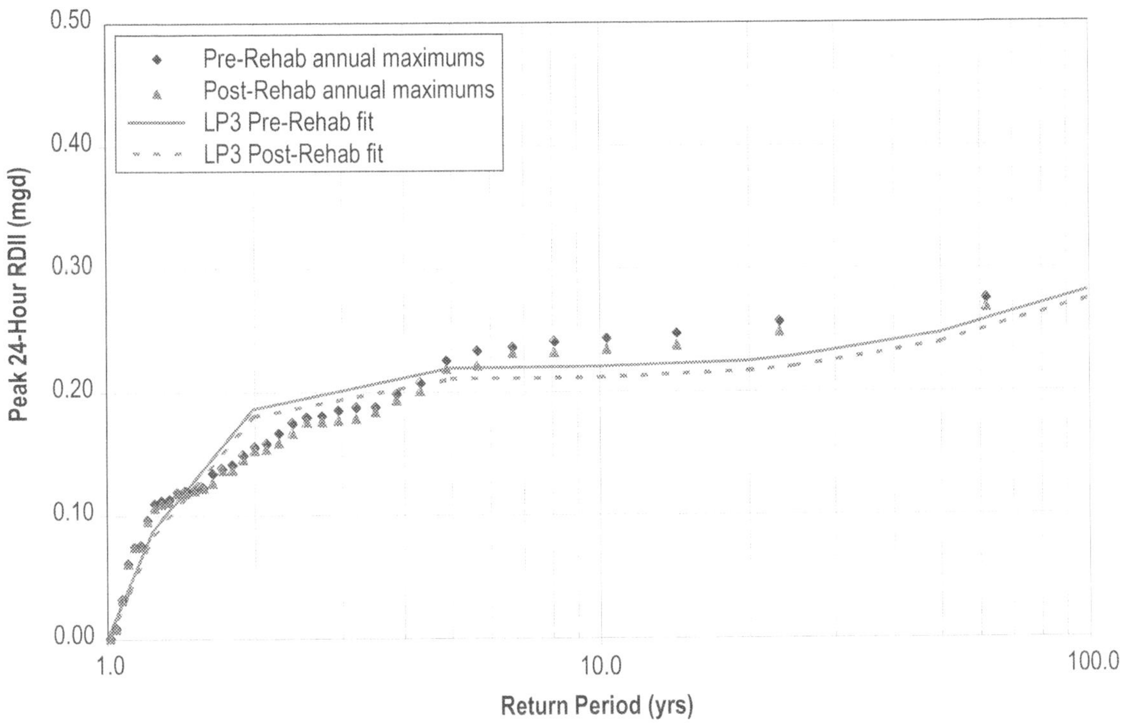

Figure 5-7. Return Period for Peak-Day (24-Hour Average) RDII for Honolulu Basin A.

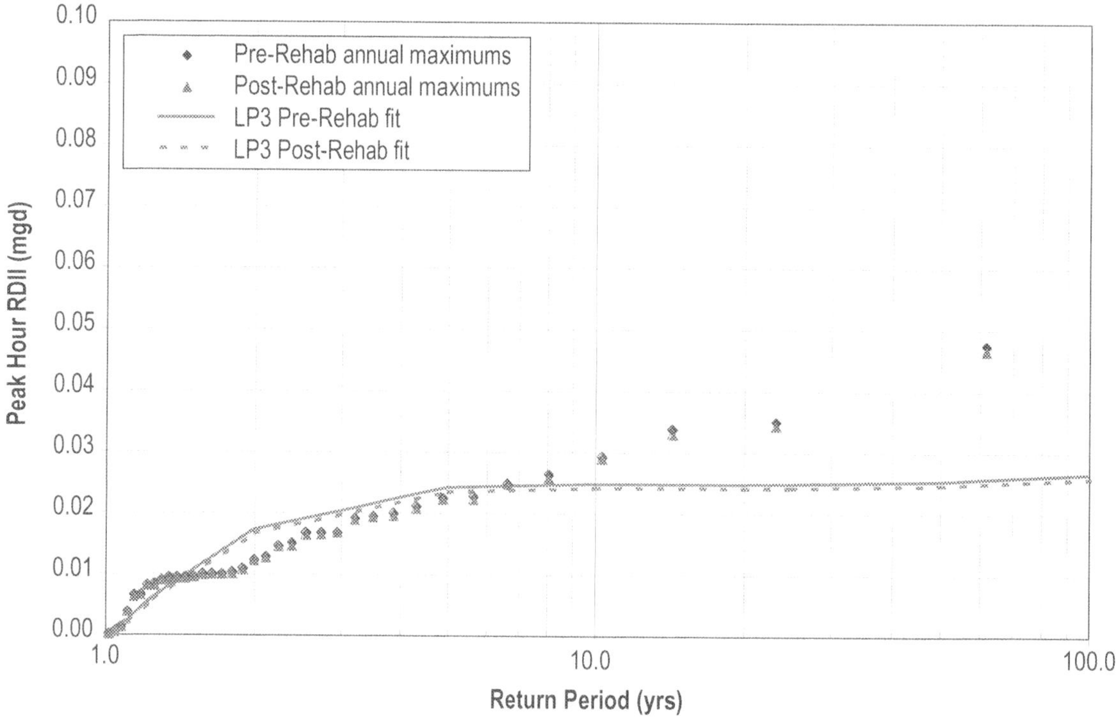

Figure 5-8. Return Period for Peak-Month (30-Day Average) RDII for Honolulu Basin A.

Basin B. The model calibrations, which Appendix B shows, indicated that there was no detectable difference between pre- and post-rehabilitation conditions. The model developed for the pre-rehabilitation period fits the post-rehabilitation data equally well. The conclusion, therefore, is that there was essentially no peak flow reduction for this project—which is supported by review of the peak measured RDII to antecedent rainfall plot (see Figure B-20 in Appendix B).

5.2.5.3 Cost-Effectiveness Analysis

Basin A. Table 5-13 summarizes the cost-effectiveness analysis performed for Basin A.

Table 5-13. Peak RDII Removal Cost-Effectiveness Measures for Honolulu Basin A.

RDII Condition	*RDII Removal Cost-Effectiveness, $/gpd*[a,b]		
	Peak Hour	*Peak Day*	*Peak Month*
5-year	$112.40	$303.09	$5,191.21
10-year	$95.87	$301.56	$4,927.26
20-year	$89.40	$320.96	$4,915.23

[a]Construction costs only.
[b]Costs are adjusted to 2002 dollars.

Basin B. The modeling indicated that there was essentially no peak RDII reduction in Basin B. No cost-effectiveness table, therefore, was prepared.

5.2.6 Conclusions

In Basin A, the sewer main rehabilitation approach was estimated to be only slightly effective in reducing RDII. Within the accuracy of the analysis, it is unlikely that any reduction occurred. The presumption is that RDII entered the collection system through defects in the private house laterals, which were not rehabilitated.

In Basin B, the spot repair approach was not effective in reducing RDII. Even though 45%, by pipe length, of the sewer mains and 11% of the manholes were rehabilitated, no consistent reduction in RDII was measured. This approach has typically not been successful on the United States mainland either because RDII enters the collection system through defects in the sewer mains and manholes that were not rehabilitated.

5.3 LOTT Case Study

The purpose of this case study analysis is to review the results of the RDII reduction study for the LOTT Wastewater Partnership basins. Three wastewater basin analyses are included: Basin OL22 Phase I, which involved sewer main replacement in the PROW; Basin OL22 Phase II, which involved replacement of private house laterals; and Basin OL24, which involved replacement of private house laterals only.

5.3.1 Project Background

LOTT is an acronym for an association of four agencies that provide regional wastewater treatment and conveyance (i.e., the cities of Lacey, Olympia, and Tumwater and the county of Thurston) in the south Puget Sound region of Washington State. As part of the 1995 Intergovernmental Contract for RDII Management and New Capacity Planning, the city of Olympia agreed to implement the plan outlined in LOTT's 1994 RDII study. That study identified basins OL22 and OL24 as areas where RDII removal would be cost effective.

Table 5-14 summarizes the characteristics of both study basins.

Table 5-14. Characteristics of LOTT Study Basins.

Basin	*Area (acres)*	*Population*	*Sewer System (feet) 8-Inch, 12-Inch, and 18-Inch Diameter*	*House Laterals*	*Sanitary Sewers (idm)*	*Number of Manholes*
OL22	130	1,036	17,517	268	28.91	48
OL24	531	1,838	28,899	292	43.39	Unknown

Basin OL22. LOTT Basin OL22 is located on the west side of Olympia, near Capitol Lake. The basin comprises 130 acres with single-family residential land use and 268 separate connections (house laterals) to the public sewer.

Figure 5-9 shows the location of the basin within Olympia, and Figure 5-10 shows a schematic of the sewer mains. Under contract with LOTT, the ADS Company installed a flow meter at the outlet of the basin as an initial RDII investigation step.

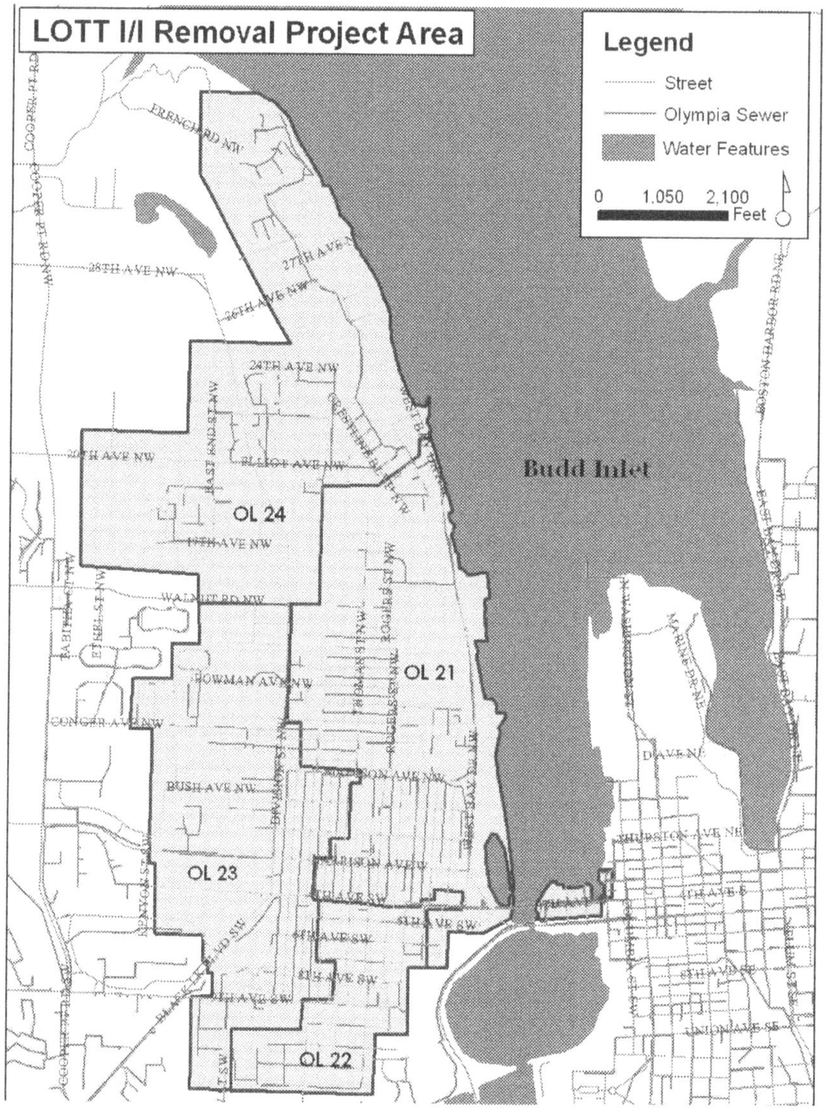

Figure 5-9. Location of LOTT Basins OL22 and OL24 within the City of Olympia.

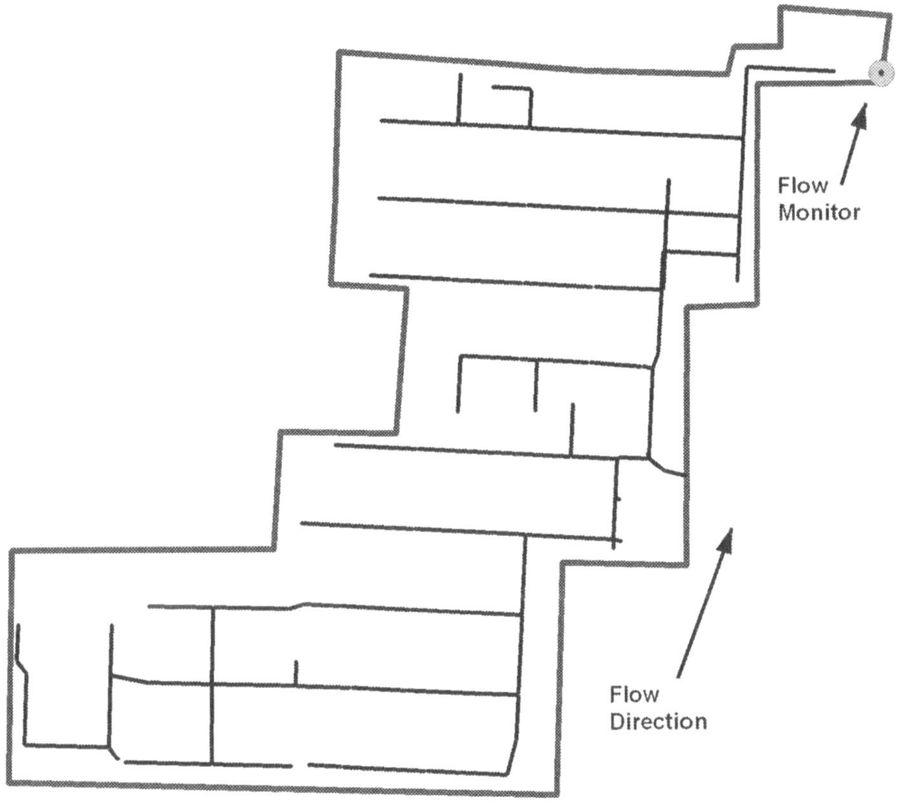

Figure 5-10. Schematic View of LOTT Basin OL22 Sewers and Drainage.

The city of Olympia undertook a program to reduce RDII by replacing sewers and manholes with the 1997 Removal Project in Basin OL22. The reduction effort in Basin OL22 was performed in two phases: the first was limited to sewers located in the PROW and the second involved house laterals on private property. The program was separated into two phases and analyzed separately to focus on these questions: by what means do extraneous flows enter the sewer system and what costs are involved in sewer rehabilitation? Answering these questions helps determine the most effective program to reduce RDII in the collection system.

Basin OL24. Basin OL24 is located along the west side of Budd Inlet, near Elliot Avenue and Crestline Boulevard, to the north of Basin OL22. The rehabilitation of Basin OL24 replaced only house laterals from the residence to the property line. Neither the sewer mains nor lower house laterals were replaced.

5.3.2 Monitoring Data

This section describes the flow and rainfall data collected as part of this sewer rehabilitation project. LOTT monitors sewer flows and rainfall with a network of gauges throughout the service area.

5.3.2.1 Flow Data

The flow meter installed in Basin OL22 as an initial step of RDII investigation in the basin was a model 1500, which is an open channel meter provided by the ADS Company. This meter uses a continuous wave Doppler for velocity measurements and ultrasonics for depth measurements. Data was stored on a 15-minute time step after being retrieved remotely through a telephone connection. The data downloads occurred on a near weekly basis. Once the data was downloaded, the standard protocol called for reviewing velocity and depth data to determine if the meters required any field attention (such as sensor scrubbing).

5.3.2.2 Rainfall Data

Rainfall records dating back to 1955 are available from the Olympia Municipal Airport weather station. Hourly rainfall data from the NOAA gauge at the Olympia Municipal Airport weather station (ID 456114) was used for model calibration and to perform long-term simulations. Although other gauges in the area are closer to Basin OL22 than the airport gauge, that gauge has the longest continuous record set. This data was used for both calibration and long-term simulation to avoid systematic differences in rain gauge performance and localized rainfall patterns. This was done to avoid bias that could occur in the RDII model results if one gauge was used to calibrate and another used for the simulations. A review of the airport gauge against the other gauges indicated that there were no significant differences during the major events used for calibration.

5.3.3 SSES Information

The city of Olympia engaged the services of a private sewer inspection contractor to provide CCTV inspection of the lateral sewers for Basin OL22. The television field work was conducted under two separate contracts, the first covering 5th and 6th Avenues and the second covering 7th, 8th, and 9th Avenues. Televised inspection was limited to the homes for which the lower house lateral was replaced in Phase I. The CCTV data was combined with information on sewer age to determine which private house laterals to replace. Ultimately, the decision was made to replace all upper house laterals that were not plastic, regardless of age. That decision accounted for 165 of 268 connections in Basin OL22.

CCTV inspection was also conducted in Basin OL24. The inspections were used to confirm the construction materials of the laterals.

5.3.4 Project Details

This section provides details on the work performed during this sewer rehabilitation project, including a discussion of project costs.

5.3.4.1 Construction Details

Basin OL22 Phase I. This part of Basin OL22's RDII project consisted of replacing the majority of sewer mains and lower house laterals, the portion of the house lateral located in the PROW to the main.

Table 5-15 presents the key elements of the rehabilitation project.

Table 5-15. LOTT Basin OL22 Phase I Construction Details.

Rehabilitation Method	*Replacement of Sewer Mains, Manhole, and Lower House Laterals with PVC Pipe*
Total Length of Pipe in Basin	17,517′
Range of Pipe Sizes	8–12″ diameter
Length of Pipe Replaced	10,512′
Number of House Laterals Replaced	195 (lower house laterals only)

Basin OL22 Phase II. This part of the RDII pilot analysis consisted of replacing most of the upper house laterals (all of which were not currently PVC pipe) in Basin OL22, the portion of the house laterals on private property. There were165 upper house laterals, measuring 7,452 LF, that were replaced with 4″ diameter high density polyethylene (HDPE) pipe during the summer of 1999. There was 210 LF of sewer main line in Alley Way that was replaced with 6″ diameter PVC pipe as part of Phase II. Public notification and involvement was an important element of Phase II because the construction occurred on private property. The city of Olympia sent out a two-part consent form to all residents in the basin whose house laterals were not new. The first part of the consent form asked for owners' permission to televise the upper house lateral and assess defects in the line; the second part of the consent form sought permission for sewer replacement construction. Residents' concerns focused on the effects of the proposed construction on landscaping, driveways, decks, etc. Homeowners were able to voice their opinions and concerns at a public meeting. At the completion of the public involvement process, most residents approved both portions of the consent form, although some wanted to wait for the inspection results and diagnosis before agreeing to allow sewer replacement.

Table 5-16 summarizes Phase II of the OL22 RDII reduction demonstration project.

Table 5-16. LOTT Basin OL22 Phase II Construction Details.

Rehabilitation Method	*Replacement of House Laterals with HDPE Pipe by Pipe Bursting; Some Open Trench Construction*
Number of Connections in Basin	268
Number of House Laterals Replaced	165 (upper house laterals only)
Sewer Mains Replaced	210′
Length of Pipe Replaced	7,452′

In Phase II, sewers were replaced from the house to the property line where the upper house lateral was joined to the lower house lateral, using a Fernco connector. While the house lateral replacement project was taking place, several direct connections from roof drains and foundation drains to the sanitary sewer were identified. Six of the 165 homes in the program experienced basement flooding after construction. City staff returned to these residences to inspect and investigate the problem. For five of the six homes, fixing the connection between the foundation drain and storm sewer solved the problem. For the other home, connecting to the storm sewer was not feasible. The foundation drain, therefore, was connected to the sanitary line and the city furnished the resident with a sump pump and generator.

Basin OL24. In this basin, the city decided to focus rehabilitation efforts on the house laterals that were expected to be the largest contributors of RDII. The basin includes an older area along the shore of Budd Inlet and a newer area where construction is recent and PVC pipe was used. Basin OL24 has 334 sewer customers connecting to the sewer system, 81 of which were found to have either concrete or fiber house laterals from examination of construction records. Eleven of these residents declined to participate in the program; one sewer was replaced as a demonstration before the project started; and five sewers were found to be cast iron or acrylonitrile-butadiene-styrene. A total of 64 house laterals, therefore, were replaced in the summer of 1998. Pipe bursting was the method agreed on for replacement. This technique begins when the leading "bursting head" expands and breaks the existing sewer, and a new HDPE pipe is dragged in behind. The method was chosen both for its limited impact on landscaping and elimination of the need for testing pipe joints.

Table 5-17 summarizes project data.

Table 5-17. LOTT Basin OL24 Construction Details.

Rehabilitation Method	*Replacement of House Laterals with HDPE Pipe by Pipe Bursting; Some Open Trench Construction*
Number of Connections in Basin	334
Number of House Laterals Replaced	64 (all concrete and fiber sewer)
Sewer Mains Replaced	None
Length of Pipe Replaced	4,290′

5.3.4.2 Construction Costs

Table 5-18 presents cost information for the Basin OL22 Phase I and Phase II projects and the Basin OL24 project.

Table 5-18. LOTT Construction Cost Summary.

Basin	*Construction Cost*[a]	*Costs Adjusted to 2002 Dollars*
OL 22 Phase I	$1,730,000	$1,953,936[a]
OL 22 Phase II	$823,000	$893,486[b]
OL24 Total	$2,553,000	$2,847,400[b]
OL24	$234,200	$254,259

[a]Costs were adjusted according to the 20-city average ENR construction cost index (ENR 5825).
[b]Costs were adjusted according to the 20-city average ENR construction cost index (ENR 6060).

Highlights of the non-project costs for these basins are as follows:

- Non-project costs for SSES activities, contract administration, engineering, and public involvement amounted to approximately 30% of the construction cost.
- The Basin OL24 non-project cost includes $145,000 for engineering and administration (62 percent of construction) and $125,000 for public involvement (54% of construction). Public involvement and engineering costs were high as this was a demonstration project and participation was voluntary. For each property, individual designs were created and reviewed with owners several times before construction. Construction inspection activities were intensive to mitigate owners' concern over damage to landscaping.

5.3.5 Post-Construction Data Evaluation

This section discusses the analysis performed before and after construction conditions. The discussion includes a review of data quality issues and an overall summary of RDII reduction success.

5.3.5.1 Calibration

Basin OL22. Flow and rainfall data was gathered and measured in Basin OL22, and sanitary base flow was projected using population estimates and unit generation rates. Brown and Caldwell's hydrologic RDII model was used to estimate the amount of extraneous water entering the sewer system. The hydrologic models made use of several equations and model parameters to simulate the movement of water in the basin. In general, the values for several of the model parameters, such as a soil's hydraulic conductivity, is not well known. The values for these parameters are determined by calibrating the model to the measured flow data. Once the model is calibrated, long-term simulations can be performed with a more widely varying set of rainfall conditions than experienced during the calibration period alone. After running the model with a long rainfall time series, RDII projections are analyzed using a standard hydrological technique that consists of fitting an LP3 distribution to the maximum peak RDII for each year. The calibration plots in Appendix C show example calibration plots for Basin OL22, post rehabilitation. The time period in Figure C-1 extends over several storms, demonstrating that the model accurately reproduces the storm peaks and RDII flow recession that occur after rainfall ends. Figure C-2 shows a more detailed view of the largest storm RDII of the preceding wet season, which occurred on December 16, 1999.

Basin OL24. Brown and Caldwell applied the method of analysis described above to estimate the reduction in RDII because of the house lateral replacement demonstration project in Basin OL24. Model calibrations were developed separately for the pre- and post-construction periods. Figures C-3 and C-4 in Appendix C show examples of the calibration plots.

5.3.5.2 Results

Basin OL22. A lengthy time series of sewer RDII was computed and analyzed. Tables 5-19, 5-20, and 5-21 present the pre- and post-construction peak hourly, peak daily, and peak monthly RDII conditions for Basin OL22.

Table 5-19. Pre-and Post-Construction Peak-Hour RDII Conditions for LOTT Basin OL22.

Condition [Row ID]	5-Year Peak-Hour RDII			10-Year Peak-Hour RDII			20-Year Peak-Hour RDII		
	mgd	gpad	gpd/idm	mgd	gpad	gpd/idm	mgd	gpad	gpd/idm
Pre-Construction [A]	2.476	19,045	85,640	3.016	23,201	104,328	3.511	27,011	121,462
Post-Construction Phase I [B]	2.055	15,807	71,078	2.517	19,360	87,057	2.921	22,466	101,023
Post-Construction Phase II [C]	0.965	7,423	33,379	0.981	7,546	33,933	0.982	7,554	33,967
RDII Removal Effectiveness Phase [D = A-B]	0.421	3,238	14,561	0.499	3,841	17,271	0.591	4,545	20,439
Additional RDII Removal Effectiveness Phase II [E = C-B]	1.090	8,384	37,699	1.536	11,814	53,124	1.939	14,912	67,056
Percent Removal Effectiveness Phase I [F = D/A]	17%			17%			17%		
Percent Additional Removal Effectiveness Phase II [G = E/A]	53%			61%			66%		
Overall Percent Removal Effectiveness [H = (D+E)/A]	70%			78%			83%		

Table 5-20. Pre-and Post-Construction Peak-Day RDII Conditions for LOTT Basin OL22.

Condition [Row ID]	5-Year Peak-Day RDII			10-Year Peak-Day RDII			20-Year Peak-Day RDII		
	mgd	gpad	gpd/idm	mgd	gpad	gpd/idm	mgd	gpad	gpd/idm
Pre-Construction [A]	1.567	12,051	54,191	1.866	14,357	64,559	2.131	16,392	73,712
Post-Construction Phase I [B]	1.113	8,562	38,500	1.330	10,228	45,991	1.487	11,439	51,437
Post-Construction Phase II [C]	0.741	5,702	25,641	0.809	6,220	27,971	0.851	6,544	29,425
RDII Removal Effectiveness Phase I [D = A-B]	0.454	3,489	15,691	0.537	4,129	18,569	0.644	4,954	22,275
Additional RDII Removal Effectiveness Phase II [E = C-B]	0.372	2,860	12,859	0.521	4,007	18,020	0.636	4,895	22,012
Overall Percent Removal Effectiveness [F = D/A]	29%			29%			30%		
Percent Additional Removal Effectiveness Phase II [G = E/A]	33%			39%			43%		
Overall Percent Removal Effectiveness [H = (D+E)/A]	62%			68%			73%		

Table 5-21. Pre-and Post-Construction Peak-Month RDII Conditions for LOTT Basin OL22.

Condition [Row ID]	5-Year Peak-Month RDII			10-Year Peak-Month RDII			20-Year Peak-Month RDII		
	mgd	gpad	gpd/idm	mgd	gpad	gpd/idm	mgd	gpad	gpd/idm
Pre-Construction [A]	0.291	2,242	10,081	0.326	2,505	11,267	0.355	2,733	12,289
Post-Construction Phase I [B]	0.105	804	3,616	0.116	890	4,000	0.121	933	4,194
Post-Construction Phase II [C]	0.148	1,137	5,113	0.150	1,155	5,193	0.150	1,157	5,205
RDII Removal Effectiveness Phase I [D = A-B]	0.187	1,438	6,465	0.210	1,616	7,266	0.234	1,800	8,095
Additional RDII Removal Effectiveness Phase II [E = C-B]	-0.043	-333	-1,497	-0.034	-265	-1,193	-0.029	-224	-1,011
Overall Percent Removal Effectiveness [F = D/A][a]	64% (49)			64% (54)			66% (58)		

[a]Statistical analysis suggests the maximum month average RDII flows were higher after Phase I and Phase II. This anomaly is assumed to result from small errors in the modeling of the two periods. Expected overall performance was estimated from Phase I results, and corresponding removals using Phase II results are shown in parentheses.

Figures 5-11, 5-12, and 5-13 show the peak RDII response (fitted to an LP3 distribution) versus recurrence interval for Basin OL22 Phase I and Phase II construction, as well as the pre-rehabilitation condition.

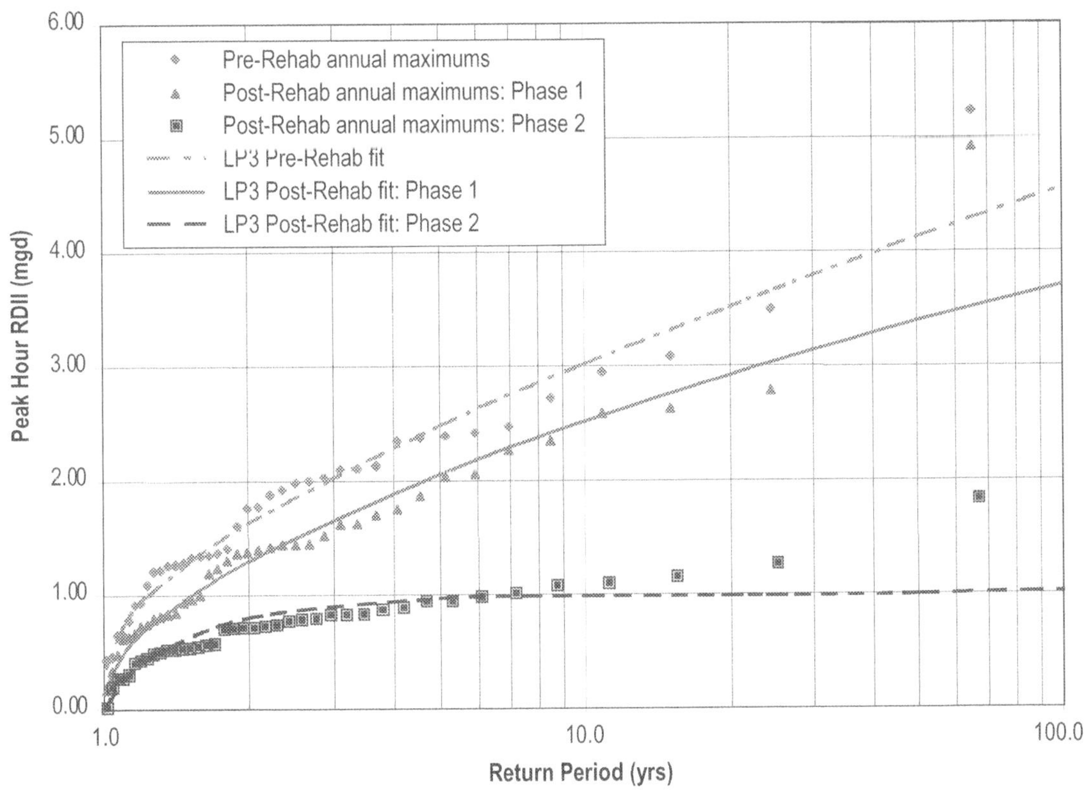

Figure 5-11. Return Period for Peak-Hour RDII for LOTT Basin OL22.

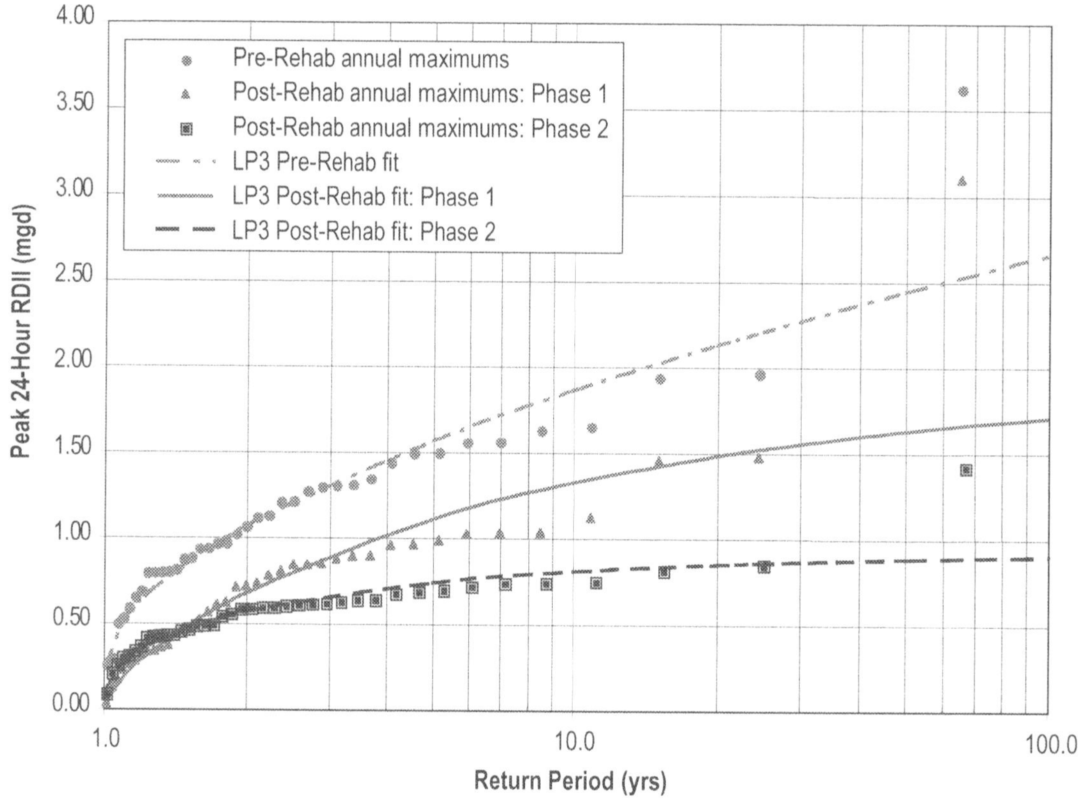

Figure 5-12. Return Period for Peak-Day (24-Hour Average) RDII for LOTT Basin OL22.

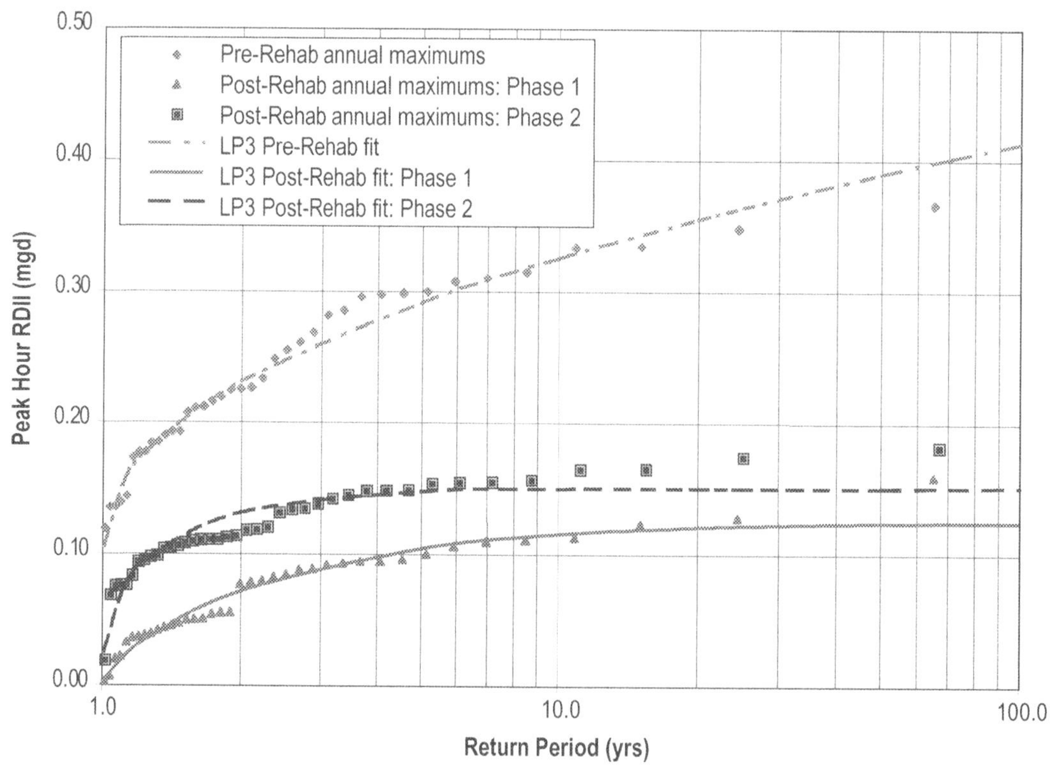

Figure 5-13. Return Period for Peak-Month (30-Day Average) RDII for LOTT Basin OL22.

Phase I activities reduced the five-year peak-hour RDII approximately 17%, the peak-day RDII approximately 29%, and the peak-month RDII 64%. Using Phase I results as the baseline, Phase II activities additionally reduced the peak-hour RDII approximately 53% and the peak-day RDII approximately 33%, and appeared to increase the peak-month RDII. The peak RDII reductions are supported by comparison of measured values to antecedent rainfall (see Figure C-6 in Appendix C).

The anomalous increase in peak-month RDII in Phase II is believed to result from small errors in the modeling or statistical technique. The corresponding return period analysis for maximum month total flow (sanitary plus RDII) shows a 12% reduction in peak-month flows after Phase II.

The results indicate that replacing sewer mains and lower house laterals significantly (49–64%) reduces the slow response, maximum month RDII. Conversely, replacing the upper house laterals reduced the fast response, peak-hour, and peak-day RDII approximately 53% and 33%, respectively, relative to the post-Phase I rehabilitation condition. Rehabilitating the private sewers did not measurably lower the maximum month RDII beyond the reduction achieved by replacing the mains and lower house laterals. These results support the idea that rapidly responding RDII enters the sewer through direct connections and rapid infiltration to the shallower upper house laterals, while the attenuated response that drives monthly RDII results from the prolonged elevated groundwater table's infiltrating the deeper sewer mains.

Basin OL24. Using the over 40-year time series precipitation record from the Olympia Municipal Airport as input to the RDII model, long-term simulations of RDII response in Basin OL24 were run. Tables 5-22, 5-23, and 5-24 present the pre- and post-construction peak hourly, peak daily, and peak monthly RDII conditions for Basin OL24.

Table 5-22. Pre- and Post-Construction Peak-Hour RDII Conditions for LOTT Basin OL24.

Condition	*5-Year Peak-Hour RDII*			*10-Year Peak-Hour RDII*			*20-Year Peak-Hour RDII*		
	mgd	*gpad*	*gpd/idm*	*mgd*	*gpad*	*gpd/idm*	*mgd*	*gpad*	*gpd/idm*
Pre-Construction	0.647	1,219	14,916	0.664	1,250	15,293	0.667	1,256	15,367
Post-Construction	0.305	573	7,018	0.309	581	7,113	0.309	582	7,121
RDII Removal Effectiveness	0.343	645	7,897	0.355	668	8,180	0.358	674	8,246
Percent Removal Effectiveness	53%			53%			54%		

Table 5-23. Pre- and Post-Construction Peak-Day RDII Conditions for LOTT Basin OL24.

Condition	5-Year Peak-Day RDII			10-Year Peak-Day RDII			20-Year Peak-Day RDII		
	mgd	gpad	gpd/idm	mgd	gpad	gpd/idm	mgd	gpad	gpd/idm
Pre-Construction	0.448	844	10,334	0.469	883	10,803	0.477	898	10,993
Post-Construction	0.200	376	4,598	0.206	387	4,738	0.208	391	4,785
RDII Removal Effectiveness	0.249	469	5,736	0.263	496	6,065	0.269	507	6,208
Percent Removal Effectiveness	56%			56%			56%		

Table 5-24. Pre- and Post-Construction Peak-Month RDII Conditions for LOTT Basin OL24.

Condition	5-Year Peak-Month RDII			10-Year Peak-Month RDII			20-Year Peak-Month RDII		
	mgd	gpad	gpd/idm	mgd	gpad	gpd/idm	mgd	gpad	gpd/idm
Pre-Construction	0.165	311	3,804	0.168	316	3,872	0.169	318	3,892
Post-Construction	0.076	143	1,744	0.077	146	1,783	0.078	147	1,800
RDII Removal Effectiveness	0.089	168	2,060	0.091	171	2,089	0.091	171	2,092
Percent Removal Effectiveness	54%			54%			54%		

Figures 5-14, 5-15, and 5-16 show the return frequency of storms before and after construction fitted to an LP3 distribution.

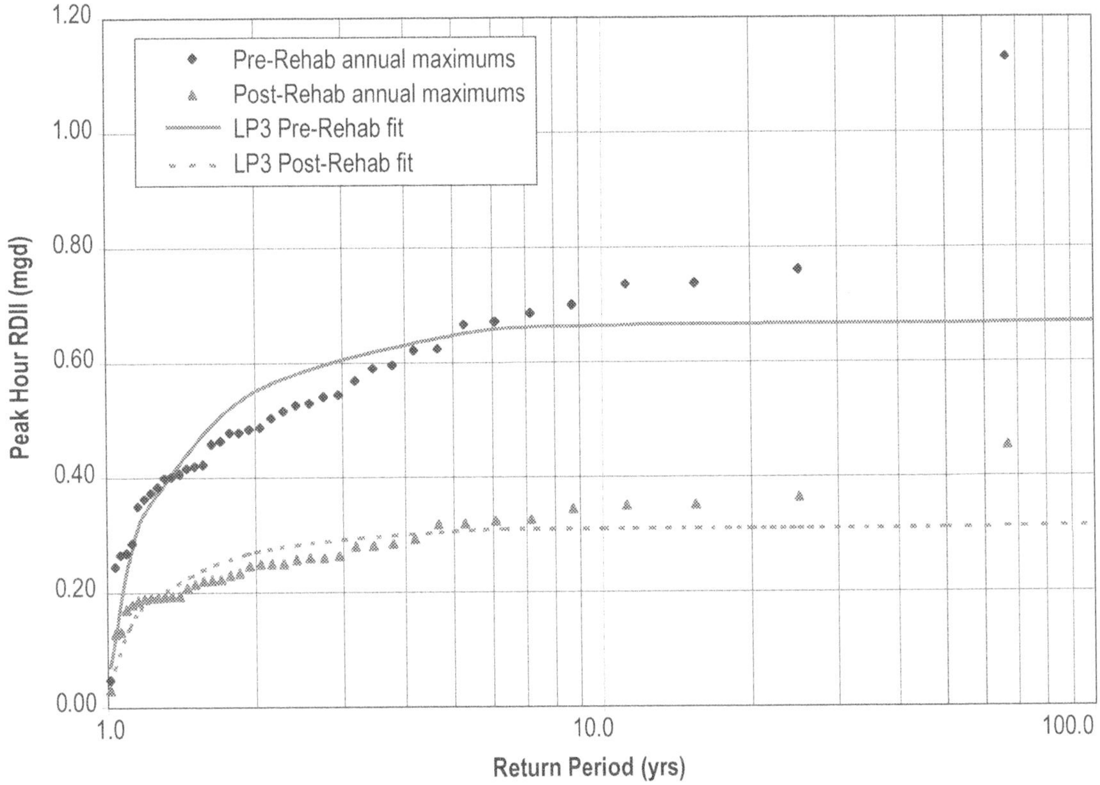

Figure 5-14. Return Period for Peak-Hour RDII for LOTT Basin OL24.

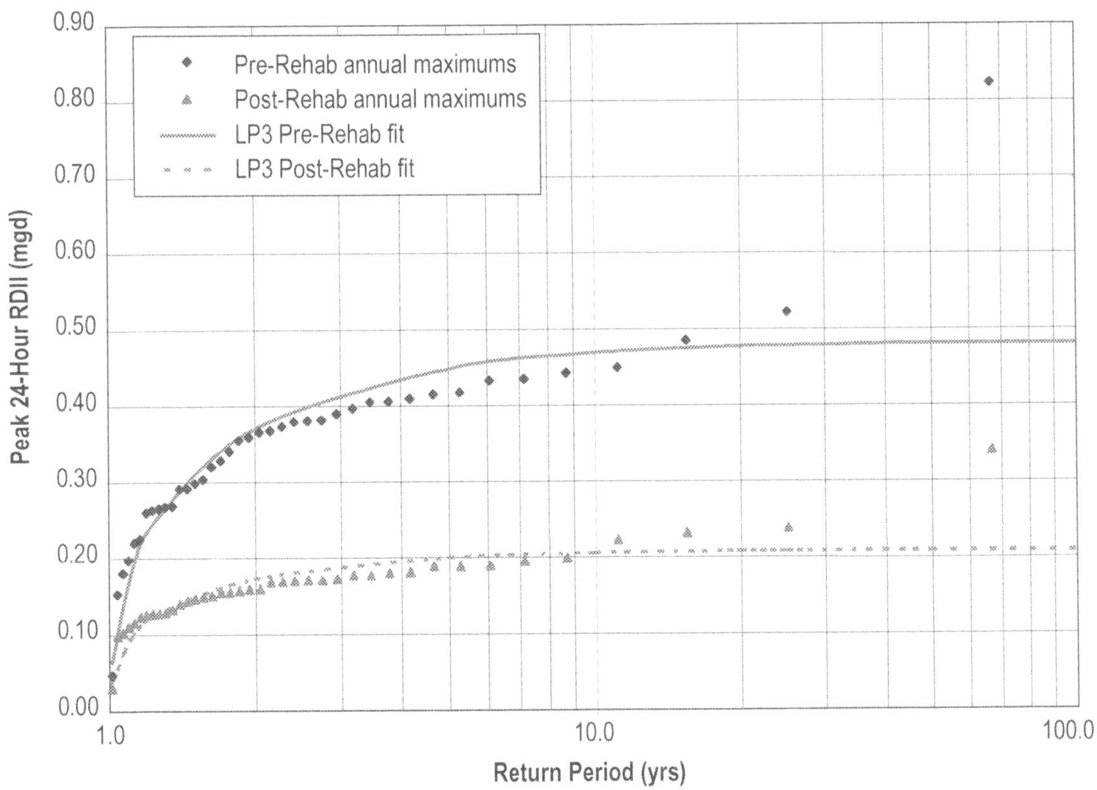

Figure 5-15. Return Period for Peak-Day (24-Hour Average) RDII for LOTT Basin OL24.

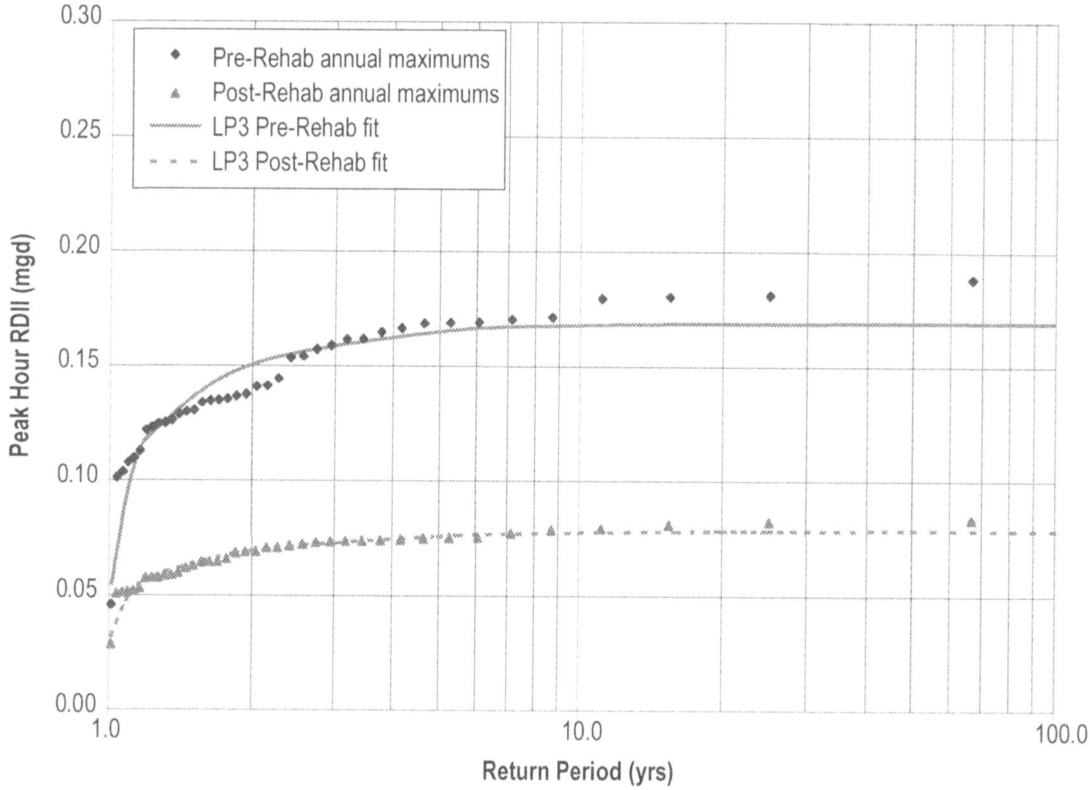

Figure 5-16. Return Period for Peak-Month (30-Day Average) RDII for LOTT Basin OL24.

The peak RDII analysis shows that the peak hour, peak 24-hour, and maximum month RDII were reduced by approximately 53% for the 10-year RDII flow event. Comparison of measured values to antecedent rainfall, which Figure C-7 shows in Appendix C, supports the peak-hour RDII reduction.

The field work performed before the rehabilitation helped to identify which house laterals had the most leaks and target them for replacement. This approach, which comprised fixing only 20% of the house laterals, removed half the RDII.

5.3.5.3 Cost-Effectiveness Analysis

Basin OL22. Tables 5-25 (Phase I) and 5-26 (Phase II) summarize the cost-effectiveness of peak RDII removal for Basin OL22.

Table 5-25. Peak RDII Removal Cost-Effectiveness Measures for LOTT Basin OL22, Phase I.

RDII Condition	*RDII Removal Cost-Effectiveness, \$/gpd*[a,b]		
	Peak Hour	*Peak Day*	*Peak Month*
5-year	\$4.64	\$4.31	\$10.46
10-year	\$3.91	\$3.64	\$9.30
20-year	\$3.31	\$3.03	\$8.35

[a]Construction costs only.
[b]Costs are adjusted to 2002 dollars.

Table 5-26. Peak RDII Removal Cost-Effectiveness Measures for LOTT Basin OL22, Phase II.

RDII Condition	*RDII Removal Cost-Effectiveness, \$/gpd*[a,b]		
	Peak Hour	*Peak Day*	*Peak Month*
5-year	\$0.82	\$2.40	—
10-year	\$0.58	\$1.72	—
20-year	\$0.46	\$1.40	—

[a]Construction costs only.
[b]Costs are adjusted to 2002 dollars.

Basin OL24. Table 5-27 presents the cost-effectiveness of RDII removal for Basin OL24.

Table 5-27. Peak RDII Removal Cost-Effectiveness Measures for LOTT Basin OL24.

RDII Condition	*RDII Removal Cost-Effectiveness, \$/gpd*[a,b]		
	Peak Hour	*Peak Day*	*Peak Month*
5-year	\$0.74	\$1.02	\$2.84
10-year	\$0.72	\$0.97	\$2.81
20-year	\$0.71	\$0.94	\$2.80

[a]Construction costs only.
[b]Costs are adjusted to 2002 dollars.

5.3.6 Conclusions

The projects in LOTT's OL22 and OL24 basins both show dramatic reductions in RDII. The results show that private lateral sewer rehabilitation will reduce the RDII seen in the peak-hour RDII analysis. Reducing the RDII from quickly entering the sewer system helps with operation of the collection system by limiting surcharging and potential overflows. Reducing peak-day and peak-month RDII aids operation of the treatment plant during storms and reduces the total monthly discharge.

The RDII reduction achieved in OL22 Phase II and OL24 suggests that house laterals are a major source of peak RDII flows. In addition, the cost of rehabilitation projects can be improved by:

1. Concentrating on a) the basins that have high levels of RDII relative to the size of the basin or baseflow or b) the number of connections.
2. Using field inspections and existing construction documents to target repair on the leakiest sewers in the basin.

The OL22 and OL24 rehabilitation projects showed pipe bursting and replacement with HDPE pipe as a viable alternative to open trench construction. Pipe bursting has the advantage of limiting the amount of surface disruption and landscape improvement.

5.4 King County Case Study

This case study analysis evaluates the effectiveness of pilot programs in the cities of Kent and Issaquah. Both communities are located in King County, Washington, near the city of Seattle. These pilot programs involved the complete replacement of the piping systems for the rehabilitation basins, including the agency mains and private house laterals. A continuous simulation model developed by King County was used to determine the effectiveness of these programs.

5.4.1 Project Background

This case study summarizes the Kent and Issaquah I/I Pilot Project: Cost and Effectiveness Evaluation, which was prepared for the King County Department of Metropolitan Services (then the Municipality of Metropolitan Seattle) in July 1995 by Brown and Caldwell, and presents the results of the Kent and Issaquah pilot programs for control of RDII.

Table 5-28 summarizes the general characteristics of both study areas.

Table 5-28. Characteristics of King County Study Basins.

Basin	*Area (acres)*	*Population*	*Area Served by Storm Sewer (acres)*	*Sanitary Sewer System (feet)*		*House Laterals*	*Storm Drainage System*
				8-Inch Diameter	*10-Inch Diameter*		
Issaquah							
Rehab	9.87	157	0	1320	0	39	No
Control	17.55	214	0	2976	0	68	No
Kent							
Rehab	9.93	113	9.93	1258	0	41	Yes
Control	25.32	252	25.32	3187	337	89	Yes

Kent. The city of Kent study area comprises two basins, Kent rehabilitation and Kent control, in a residential neighborhood that is bordered by an industrial area and Highway 167. Both basins have similar characteristics; however, only the Kent rehabilitation basin had rehabilitation performed on its sewer system. The sewer mains and house laterals of both basins, which were constructed approximately 50 years ago, are made of clay. In addition, the Kent basins have a key manhole through which all sewage flows.

The Kent study area is served by a separate storm drainage system. Soils within the Kent study area are alluvial loam soils, which are poorly drained and cause a high water table. These soil conditions contribute to storm-related ponding and flooding. The topography of the study area is generally flat.

Before construction, the hourly peaking factor for the Kent rehabilitation basin was 4.7. Peaking factors are the ratios of peak-hour WWFs to ADWFs.

Figure 5-17 shows the city of Kent study area.

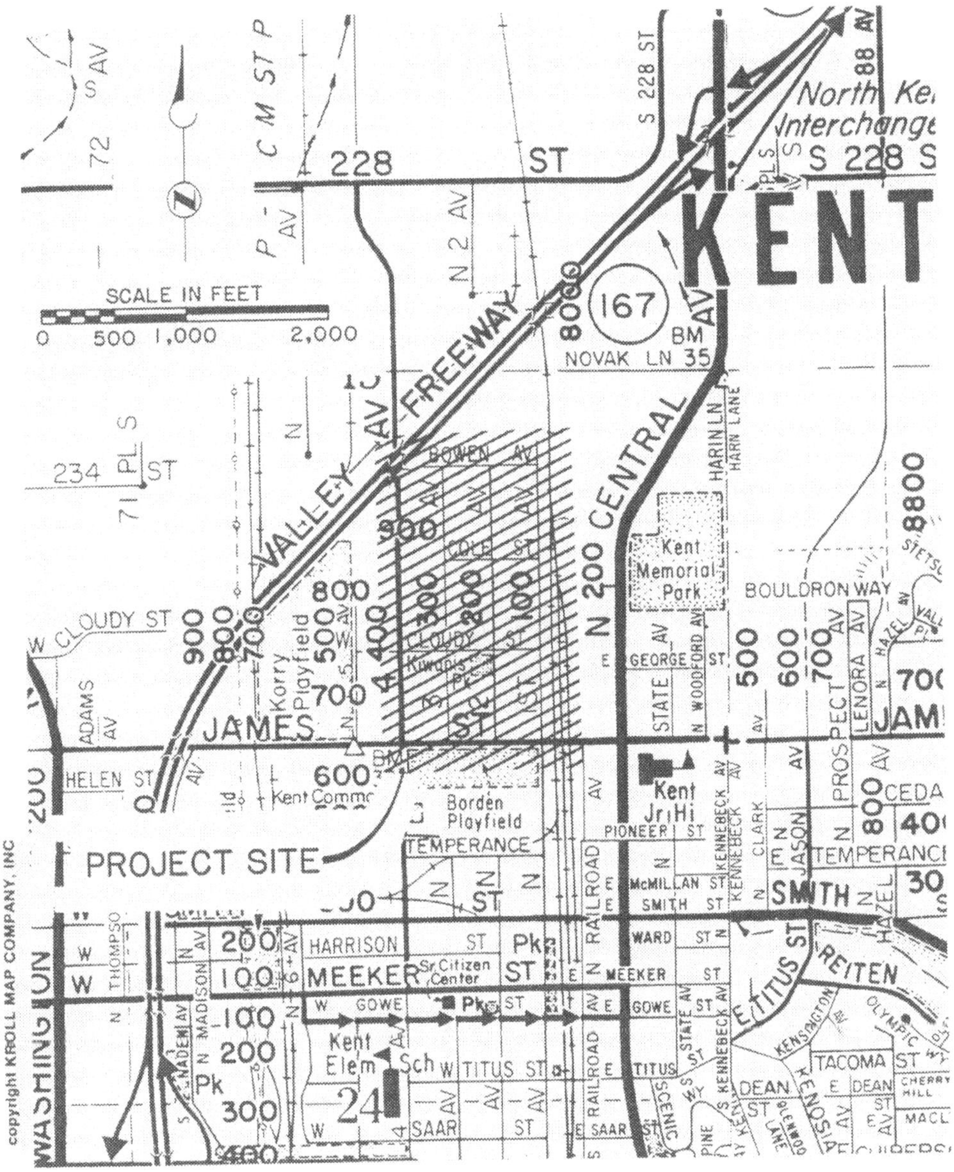

Figure 5-17. Location of Kent Study Area in King County.

Issaquah. The city of Issaquah study area comprises two basins, Issaquah rehabilitation and Issaquah Control, in an established residential neighborhood with a history of RDII problems. Both basins have similar characteristics; however, only the Issaquah rehabilitation basin had rehabilitation performed on its sewer system. The study area's sewer system was constructed in 1936 with clay pipe and oakum joint seals. Both Issaquah basins have a key manhole through which all sewage flows.

The Issaquah study area is not served by a storm drainage system. Runoff flows overland into the adjacent local storm sewer. The topography of the study area is generally flat, with soils consisting primarily of glacial outwash and alluvial loam with high permeability. These conditions provide a means for water to percolate through the soil and move laterally and vertically through the soil column. Before construction, the hourly peaking factor for the Issaquah rehabilitation basin was 3.8.

Figure 5-18 shows the city of Issaquah study area.

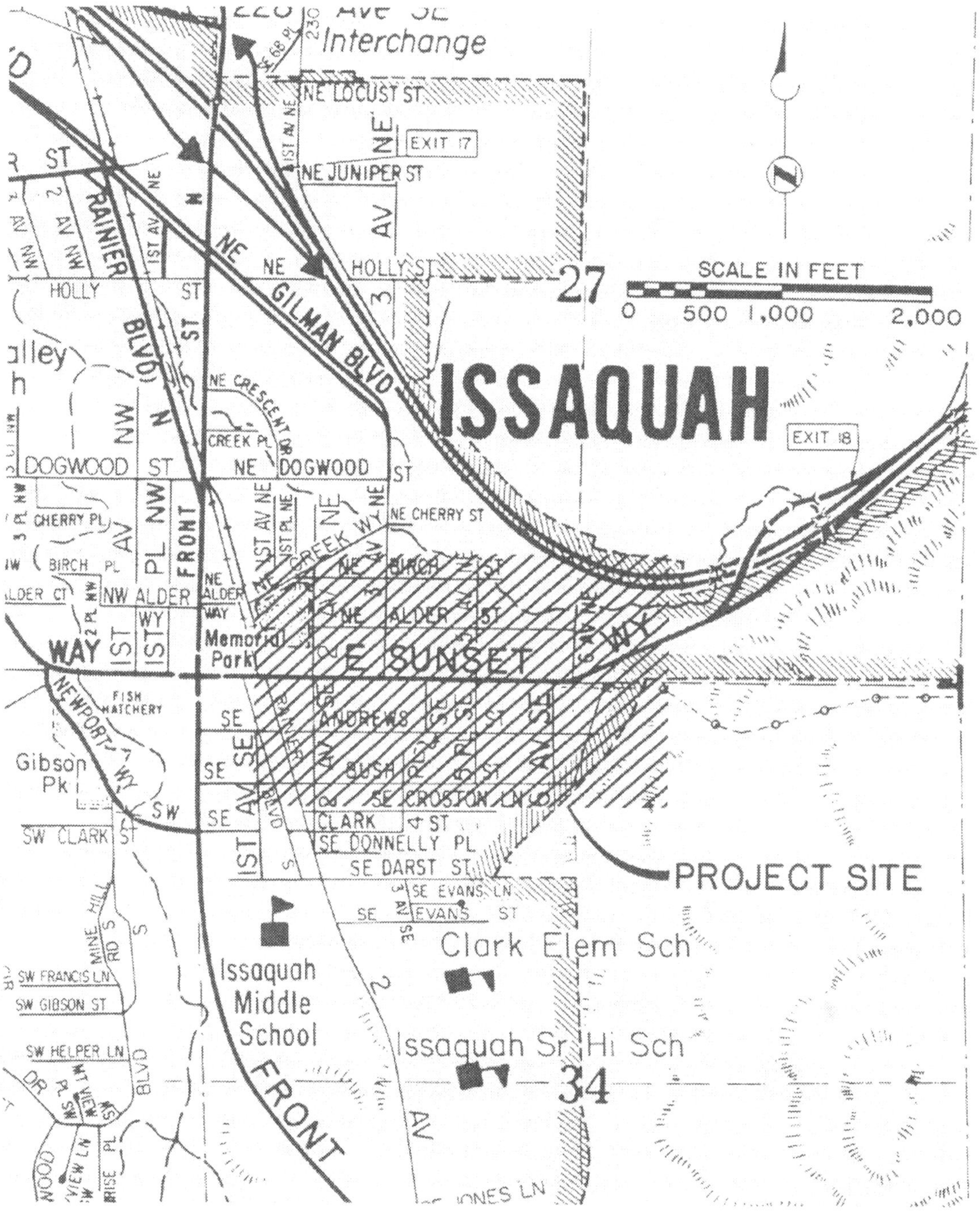

Figure 5-18. Location of Issaquah Study Area in King County.

Water Environment Research Foundation

5.4.2 Monitoring Data

This section describes the flow and rainfall data collected as part of this sewer rehabilitation project.

5.4.2.1 Flow Data

Before rehabilitation, flow monitoring for the Kent rehabilitation and Issaquah sewer systems was conducted from January–March 1988 and from October 1988–February 1989. The flow monitoring from January–March 1988 produced little usable data because the metering probes were chronically clogged, and the depth of the low flows did not cover the velocity/depth probe placed in the invert of the sewer. During the October 1988–February 1989 monitoring period, trapezoidal weirs and bubbler-type depth sensors were used to measure flow. Marsh-McBirney Flo-totes were installed in the outlet pipes from the basins for the post-rehabilitation flow monitoring. The meters monitored flow from December 1992 until at least July 1995. No data exists for a three-week period in January and February 1993 because one of the meters failed. Meters for both the pre- and post-rehabilitation periods recorded flow, depth, and velocity in five-minute intervals. It is unknown how frequently or to what extent the flow meters were calibrated and maintained.

5.4.2.2 Rainfall Data

Kent. Rain data for the Kent basins was obtained from King County gauges 03u and 41u. Gauge 03u is located two miles northeast of the Kent site, and gauge 41u is located three miles southwest of the site near Star Lake.

Issaquah. Rain data for the Issaquah basin was obtained from King County gauges 14u and 67u. Gauge 14u is located one mile northeast of the Issaquah site, and gauge 67u is located one mile west of the site.

5.4.3 SSES Information

Around 1987, LSSES work was completed, which included smoke testing and CCTV inspection of sewers. This LSSES work concluded that the Kent and Issaquah basins were suitable for rehabilitation.

Kent. In March 1989, additional television inspection of the Kent basins during light-to-moderate rainfall concluded that the most significant source of infiltration was leaky service laterals.

Issaquah. Smoke testing of the Issaquah basins indicated that three roof drains and one open backyard pipe were directly connected to the sewer system; however, no direct connections were found.

5.4.4 Project Details

This section provides details on the work performed during this sewer rehabilitation project, including a discussion of project costs.

5.4.4.1 Construction Details

The construction methods used in the Kent and Issaquah project basins consisted of replacing the sewer main with PVC pipe, replacing the manholes with new precast concrete manholes, and replacing the house laterals with PVC pipe. These methods were intended to reduce the amount of infiltration resulting from leaking sewers and manholes.

Kent. The Kent construction project consisted of replacing 1,258′ (100%) of 8″ diameter local sewer main and 41 house laterals (100%) with a total length of 3,200′, or 4.45 idms of pipe. All manholes were replaced as part of the project, which was completed on September 30, 1992.

Issaquah. The Issaquah construction project consisted of replacing 1,320′ (100%) of 8″ diameter sewer main and approximately 730 feet of 4″ and 6″ diameter house laterals. Of a total of 39 laterals, 36 (i.e., 92%) were replaced. All manholes were replaced as part of the project, which was constructed between January and March 1993.

5.4.4.2 Construction Costs

Kent. Table 5-29 summarizes the construction costs for the Kent rehabilitation basin.

Table 5-29. Kent Rehabilitation Basin in King County Construction Cost Summary.

Description	*Unit*	*Quantity*	*Unit Price*[a]	*Cost*
Mobilization/Demobilization	lump sum	1	$17,000	$17,000
Clear and Replace Landscaping	yd^2	1,300	$4.82	$6,266
Remove and Replace Pavement	ft^2	691	$2.13	$1,472
Remove and Replace Gravel Alleyway	ft^2	13,970	$0.35	$4,890
Remove and Replace Utilities		0	$0.00	$0
Replace Sewer with 8-Inch Ductile Iron Pipe (DIP)	LF	55	$61.82	$3,400
Replace Sewer with 8-Inch PVC Pipe	LF	1,258	$41.27	$51,918
Replace House Laterals	LF	2,924	$11.44	$33,451
Subtotal				$118,396
Sales tax			8.20%	$9,708
Total Construction Cost				$128,104
Total Construction Cost Adjusted to 2002[b] Dollars				$169,067

[a]Costs reflect raw dollars, unadjusted for inflation.
[b]Costs were adjusted according to the 20-city average ENR construction cost index (ENR 4985).

Issaquah. Table 5-30 summarizes the construction costs for the Issaquah rehabilitation basin.

Table 5-30. Issaquah Rehabilitation Basin in King County Construction Cost Summary.

Description	*Unit*	*Quantity*	*Unit Price*[a]	*Cost*
Mobilization/Demobilization	lump sum	1	$10,000.00	$10,000
Traffic Control	hours	256	$20.00	$5,120
Clear and Replace Landscaping	yd^2	343	$4.37	$1,499
Concrete on Private Property	1	25	$270.00	$6,750
Remove and Replace Asphalt Concrete Pavement	ft^2	1,120	$1.96	$2,195
Remove and Replace Crushed Surface	ft^2	20,336	$0.52	$10,575
Remove and Replace Utilities		0	$0.00	$0
Replace Sewer with 8-Inch DIP		0	$0.00	$0
Replace Sewer with 8-Inch PVC Pipe	LF	1,320	$59.16	$78,091
Replace House Laterals	LF	1,030	$14.72	$15,162
Change Order	LF	1	$8,370.00	$8,370
Subtotal				$137,762
Sales tax			8.20%	$11,296
Total Construction Cost				$149,058
Total Construction Cost Adjusted to 2002[b] Dollars				$188,226

[a]Costs reflect raw dollars, unadjusted for inflation.
[b]Costs were adjusted according to the 20-city average ENR construction cost index (ENR 5210).

5.4.5 Post-Construction Data Evaluation

This section discusses the analysis performed before and after construction conditions. The discussion includes a review of data quality issues and an overall summary of RDII reduction success.

King County's runoff model was used to simulate sanitary and RDII flows for both the Kent and Issaquah collection systems. This model is a physically based one that simulates base flow, dry weather infiltration, rainfall-dependent inflow, rainfall-dependent infiltration, and long-term groundwater leakage. The model uses a kinematic wave formulation in solving the St. Venant equations for overland flow and flow through the sewer pipes.

5.4.5.1 Calibration

Kent. Calibration plots comparing the measured flow to the simulated flow for the Kent basins are located in Figures D-1 through D-16 in Appendix D. In general, there is good agreement between the measured flow data and the simulated flows for the Kent basins during times when the metered data appears accurate. According to King County modeling staff, however, the sewer pipe appeared to be clogged at times, as indicated when the measured flow looks higher than the simulated flow.

Issaquah. Calibration plots comparing the measured flow to the simulated flow for the Issaquah basins are located in Figures D-17 and D-18 in Appendix D. No large storms occurred during the 1993–1994 wet season for the Issaquah basins, which made calibration difficult. In addition, there is some uncertainty in the measured flow data for the Issaquah basins because clogging appeared to take place during rainfall, as was evident when the measured flow (based on level) appears to be higher than the simulated flow. There appears to be reasonable agreement between the simulated and monitored flow during times when the metered data seems accurate.

5.4.5.2 Results

Kent. The sewer rehabilitation in the Kent rehabilitation basin resulted in a reduction of 70–80% in overall RDII. Infiltration was reduced by about 80% and peak inflow associated with the 20-year storm was reduced by 50–70%.

Table 5-31 summarizes these results. Measured flow and rainfall data was not available for comparison of peak RDII to antecedent rainfall volume.

Table 5-31. Pre- and Post-Construction Peak-Hour RDII Conditions for King County Kent Rehabilitation Basin.

Condition	*5-Year Peak-Hour RDII*			*10-Year Peak-Hour RDII*			*20-Year Peak-Hour RDII*		
	gpd	*gpad*	*gpd/idm*	*gpd*	*gpad*	*gpd/idm*	*gpd*	*gpad*	*gpd/idm*
Pre-Construction	234,437	23,680	126,041	271,453	27,420	145,943	308,470	31,159	165,844
Post-Construction	52,470	5,300	28,210	61,380	6,200	33,000	71,280	7,200	38,323
RDII Removal Effectiveness	181,967	18,380	97,832	210,073	21,220	112,943	237,190	23,959	127,521
Percent Removal Effectiveness	78%			77%			77%		

Issaquah. RDII in the Issaquah rehabilitation basin was reduced by 50–60% for the simulated peak 20-year storm flow as a result of sewer rehabilitation.

Table 5-32 summarizes the estimated reductions in peak infiltration, peak inflow, and total peak RDII associated with the 20-year storm. Measured flow and rainfall data was not available for comparison of peak RDII to antecedent rainfall volume.

Table 5-32. Pre- and Post-Construction Peak-Hour RDII Conditions for King County Issaquah Rehabilitation Basin.

Condition	5-Year Peak-Hour RDII			10-Year Peak-Hour RDII			20-Year Peak-Hour RDII		
	gpd	gpad	gpd/idm	gpd	gpad	gpd/idm	gpd	gpad	gpd/idm
Pre-Construction	67,878	6,877	32,018	73,087	7,405	34,475	78,213	7,924	36,893
Post-Construction	33,459	3,390	15,783	36,914	3,740	17,412	40,368	4,090	19,042
RDII Removal Effectiveness	34,419	3,487	16,235	36,173	3,665	17,063	37,844	3,834	17,851
Percent Removal Effectiveness	51%			49%			48%		

5.4.5.3 Cost-Effectiveness Analysis

Kent. Table 5-33 presents the peak RDII removal cost-effectiveness measures for the Kent basin.

Table 5-33. Peak RDII Removal Cost-Effectiveness Measures for King County Kent Rehabilitation Basin.

RDII Condition	Peak Hour RDII Removal Cost-Effectiveness, $/gpd[a,b]
5-year	$0.93
10-year	$0.80
20-year	$0.71

[a]Construction costs only.
[b]Costs are adjusted to 2002 dollars.

Issaquah. Table 5-34 presents the peak RDII removal cost-effectiveness measures for the Issaquah basin.

Table 5-34. Peak RDII Removal Cost-Effectiveness Measures for King County Issaquah Rehabilitation Basin.

RDII Condition	Peak Hour RDII Removal Cost-Effectiveness, $/gpd[a,b]
5-year	$5.47
10-year	$5.20
20-year	$4.97

[a]Construction costs only.
[b]Costs are adjusted to 2002 dollars.

5.4.6 Conclusions

In conclusion, the Kent and Issaquah projects removed significant amounts of RDII by replacing the agency and house laterals.

Kent. Construction on the Kent rehabilitation basin consisted of replacing 1,300′ of local sewer main and 3,200′ of house laterals, resulting in a reduction of 70–80% in peak RDII associated with the 20-year storm.

Issaquah. Construction on the Issaquah rehabilitation basin consisted of replacing 1,318′ of local sewer main and 730′ of house laterals, resulting in a reduction of 50–60% in peak RDII associated with the 20-year storm.

5.5 McCandless Case Study

The purpose of this case study analysis is to describe the effectiveness of abating RDII in three basins of the McCandless Township Sanitary Authority in Pittsburgh, Pennsylvania. This analysis will provide information on the 1999 Lowries Run Sewer System I/I Abatement Program and the results of analyzing flow and rainfall data from that study. The rehabilitation consisted of the installation of fold-and-form pipe liners in the main sewer pipe where defects were noted and the rehabilitation of public domain laterals and manholes.

5.5.1 Project Background

The McCandless authority system is located in Allegheny County, Pennsylvania. The authority selected three sub-basin systems in generally wet areas for abatement of RDII through rehabilitation of the collection system and lateral piping. These basins represent differing terrains of wetlands, streets, backyard areas, and hillsides. The study, therefore, provided a cross section of various ground conditions and groundwater tables. The land use for all the basins is 100% residential, and the pipe sizes are 8″ in diameter. The authority undertook a program to enforce ordinances that restrict illegal connections to the collector sewer system and to rehabilitate the system in the public domain.

Table 5-35 summarizes the characteristics of the study basins.

Table 5-35. Characteristics of McCandless Study Basins.

Basin	Area (acres)	Sewer System (feet) 8-Inch Diameter	House Laterals	Sanitary Sewers (idm)	Number of Manholes
Moreland Road	10.9	2,895	26	4.39	13
Woodland Road	29.6	3,116	31	4.72	15
Willoughby Road	7.4	1,900	17	2.88	14
Emerald Drive (control basin)	15.3	2,519	21	3.82	13

Moreland Road Basin. The Moreland Road Basin comprises 26 homes in an area of 10.92 acres. Sewers are located in swampy areas with a high groundwater table. The age of the sewer collection system is approximately 25 years.

Figure 5-19 shows the Moreland Road Basin sewer system (Inwood Road Flow Meter), which has 13 segments of VCPs, totaling 2,895 LF. The sewer basin has 13 manholes.

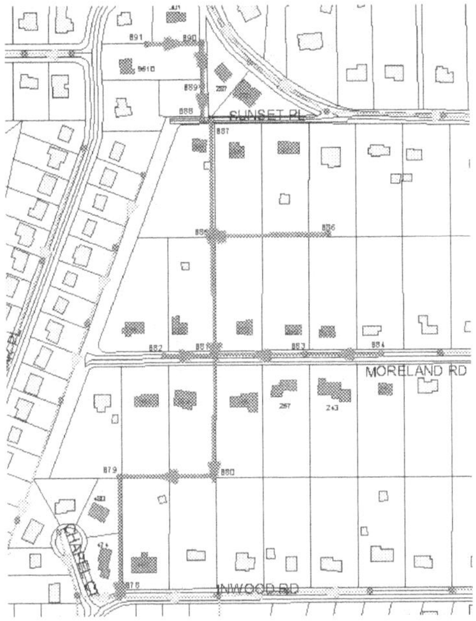

Figure 5-19. McCandless Moreland Road Basin.

Woodland Road Basin. The Woodland Road Basin comprises 31 homes in an area of 29.57 acres. Sewers are located in street areas with varying groundwater tables. The age of the sewer collection system is approximately 40 years.

Figure 5-20 shows the Woodland Road Basin sewer system (MCCLCHN Flow Meter), which has 15 manholes and 13 vitrified clay sewer reaches, totaling 3,116 LF.

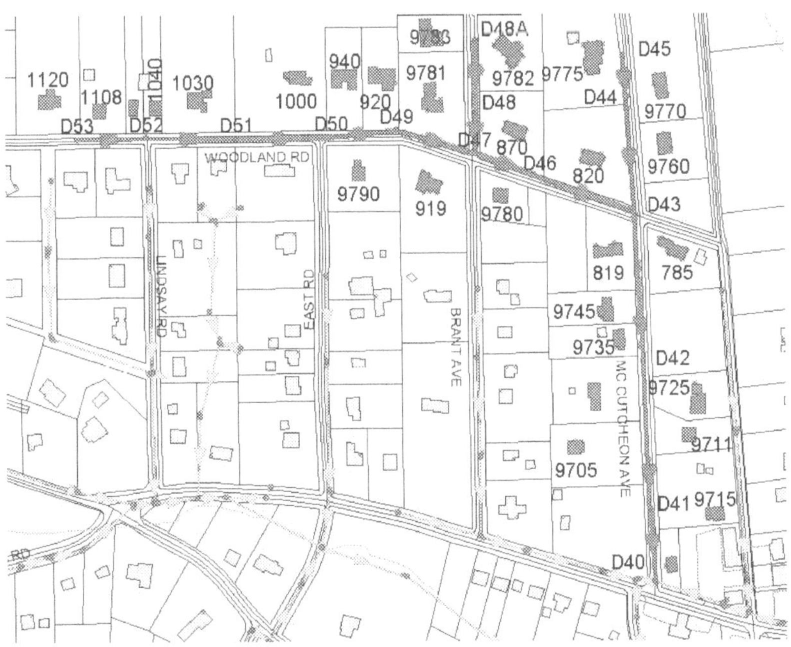

Figure 5-20. McCandless Woodland Road Basin.

Willoughby Road Basin. The Willoughby Road, Breezewood Plan No. 5 Basin comprises 7.43 acres. Sewers are located in hillside areas of varying groundwater tables. The age of the sewer collection system is approximately 30 years.

Figure 5-21 shows the Willoughby Road, Breezewood Plan No. 5 Basin sewer system (Sloop Road Flow Meter), which has 14 manholes and 13 VCP sewer reaches, totaling 1,900 LF.

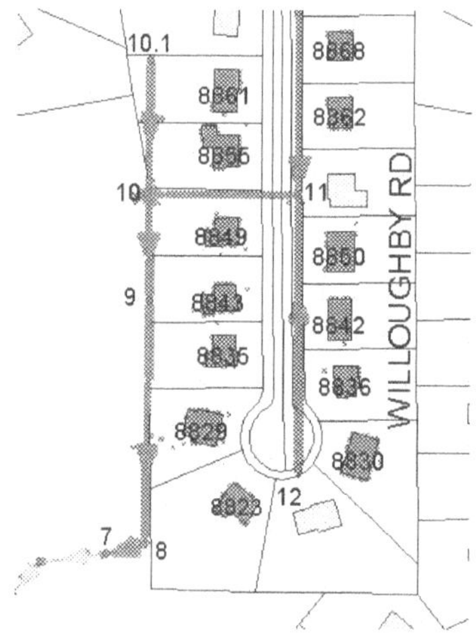

Figure 5-21. McCandless Willoughby Road Basin.

Emerald Drive Basin. A fourth basin system, the Emerald Drive Basin, was utilized as a control to compare the rates and volumes of RDII conditions experienced by the other three basin areas, before and after rehabilitation.

Figure 5-22 shows the Emerald Drive (control) Basin, which comprises 21 homes in an area of 15.33 acres. The age of the sewer collection system is approximately 25 years. The sewer system has 13 manholes and 13 vitrified clay sewer reaches, totaling 2,519 LF.

Figure 5-22. McCandless Emerald Drive (Control) Basin.

5.5.2 Monitoring Data

This section describes the flow and rainfall data collected as part of this sewer rehabilitation project.

5.5.2.1 Flow Data

The authority began a flow-monitoring program approximately three years before the RDII Abatement Program, which acted as a snapshot of the infiltration issues in the systems. Flow monitors were installed in the downstream manholes of each of the three sub-basins and control sub-basin. Flows were measured with meters using a calibrated weir and by depth measurement. Flow meters were Isco 4230 Bubble Meters (flow is measured by a weir); data was collected for review on a weekly basis for pre-construction evaluations. Flow monitoring was similarly conducted for post-constructions comparisons to evaluate the success of RDII abatement and has been ongoing. The rehabilitated basins data comparisons with the non-rehabilitated basin formed the basis of analysis.

5.5.2.2 Rainfall Data

The rain gauge used for the analysis is located approximately five miles south of the four studied areas.

5.5.3 SSES Information

The authority conducted CCTV inspections in each basin that indicated several defects. A significant number of defects was noted that involved infiltration at lateral connections to the main line sewer.

The authority conducted dye testing to determine houses that had illegal roof leader or foundation connections to the sewer system. Property owners of houses that failed the inspection were given several months to make corrections before the house was retested. In some cases, new sump pump connections to storm drains were installed; however, in other cases, roof leaders were redirected to drain on the property because of lack of better ordinance guidance.

5.5.4 Project Details

This section provides details on the work performed during this sewer rehabilitation project, including a discussion of project costs.

5.5.4.1 Construction Details

The PROWs in the basins are 20′. Each house has a dedicated private house lateral and sewer lateral in the public domain that connects to the collector sewer. Private house laterals were not rehabilitated. The authority contracted to install U Liner fold-and-form polyethylene liners in all basin collector sewers. Authority staff used open cut techniques for complete replacement of each lateral connection (approximately 8–10′) and installation of observation ports. Connections to the main line were sealed with Sika products, and all laterals were bedded completely with limestone backfill to avoid settlement.

The program of rehabilitation of the three basin systems was accomplished in the following sequence:

- Dye testing was conducted to locate possible sources of extraneous flows. If the residential rain leaders were connected, the house failed. The resident then corrected the problem and the house was reinspected.
- Flow monitoring of each basin was initiated.

- All existing collector sewer sections were rehabilitated by installation of fold-and-form pipe lining.
- Individual lateral connections were made, which consisted of the piping immediate to and including the wye connection saddle to the collector sewer. Inspection ports were installed near the street curb, and sealed couplings were installed at the point of building service sewer connection to the lateral piping. These inspection ports were installed to monitor flows within service piping from houses during heavy rainfalls.
- Flow monitoring was reconducted to evaluate the success of RDII abatement.
- All manholes were sealed with the Sauereisen "no entry" sealing system, along with the replacement of required manhole lids and/or other required repairs.
- Final flow monitoring was conducted to further evaluate the percentage reduction in RDII. Comparisons were made to the flow-monitoring data of the control system.

Moreland Road Basin. Rehabilitation was completed in September 1999. House lateral rehabilitation in the PROW of the main line to the property line was completed between June and July 2000. Roof leaders at 10 of the 26 homes in the basin were disconnected during the course of the project; these disconnections occurred between July 1998 and September 2000.

Woodland Road Basin. Rehabilitation was completed in October 1999. House laterals in the PROW of the main line were completed between April and May 2000. Roof leaders at six of 31 homes were disconnected between January 1999 and March 2001. In July 2001, all manholes in this system were rehabilitated and sealed, including interiors and castings for RDII abatement.

Willoughby Road Basin. The main line rehabilitation was completed in September 1999, and house laterals were rehabilitated to the property line between May and June 2000. No roof leader disconnects were noted.

Emerald Drive Basin. Utilized as a control, the Emerald Drive basin was not rehabilitated.

5.5.4.2 Construction Costs

The estimated total project cost was $453,000. Table 5-36 summarizes the construction costs for each basin.

Table 5-36. McCandless Construction Cost Summary.

Construction Costs	Moreland Road	Woodland Road	Willoughby Road
Equipment	$4,545.00	$6,242.31	$5,736.24
Material	$22,390.10	$18,109.83	$11,747.44
Labor	$23,354.51	$25,234.50	$22,890.82
Total	$50,289.61	$49,586.64	$40,374.50
Total Adjusted to 2002 Dollars	$54,598[a]	$53,835[a]	$42,699[b]

[a]Costs were adjusted according to the 20-city average ENR construction cost index (ENR 6060).
[b]Costs were adjusted according to the 20-city average ENR construction cost index (ENR 6221).

5.5.5 Post-Construction Data Evaluation

This section discusses the analysis performed before and after construction conditions. The discussion includes a review of data quality issues and an overall summary of RDII reduction success.

5.5.5.1 Calibration

Much of the data collected was during snowmelt conditions. In addition, the distance of the rain gauge from the basins resulted in inconsistent flow responses. As a result, there was insufficient flow-monitoring data available to perform model calibration and LP3 analyses. The post-construction data analysis, therefore, was performed by comparing peak RDII in each study basin to peak RDII in the control basin before and after rehabilitation.

5.5.5.2 Results

The analysis of the flow data for the peak RDII reduction consisted of comparison of the peak RDII in individual precipitation events from the pilot basin meters to the peak RDII in the corresponding event at the control basin meter.

Figures 5-23 through 5-25 show the peak RDII comparison plots for the three pilot basins. In theory, a reduction in peak RDII would be shown as a change in the slope of a best-fit line to the pre- and post-rehabilitation period data.

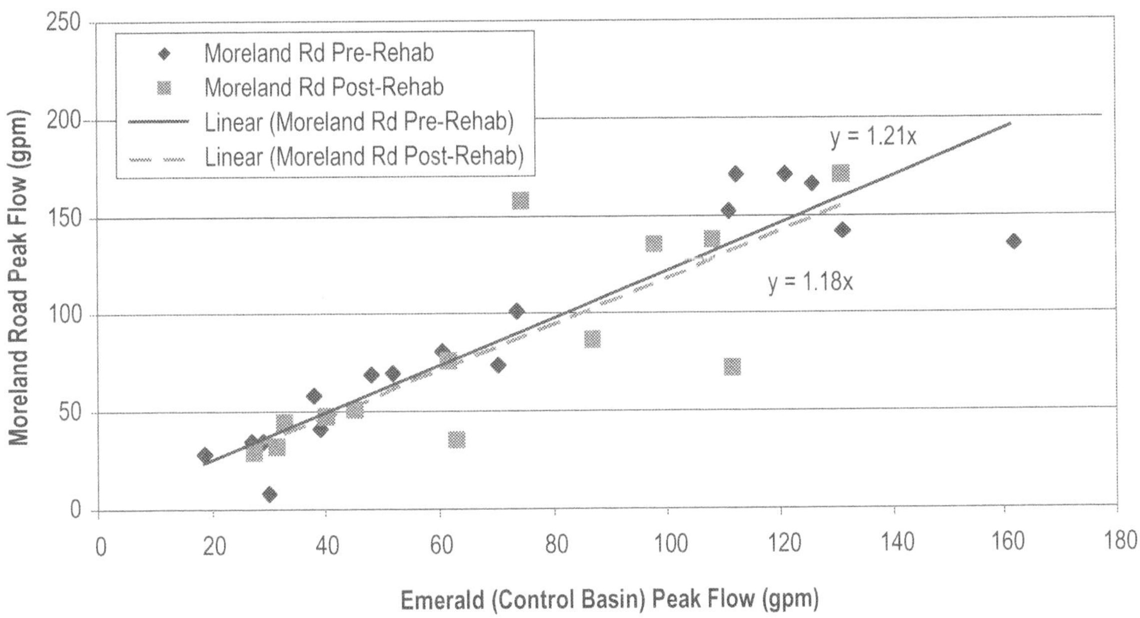

Figure 5-23. Comparison of Peak RDII for McCandless Willoughby Road Basin and Control Meter.

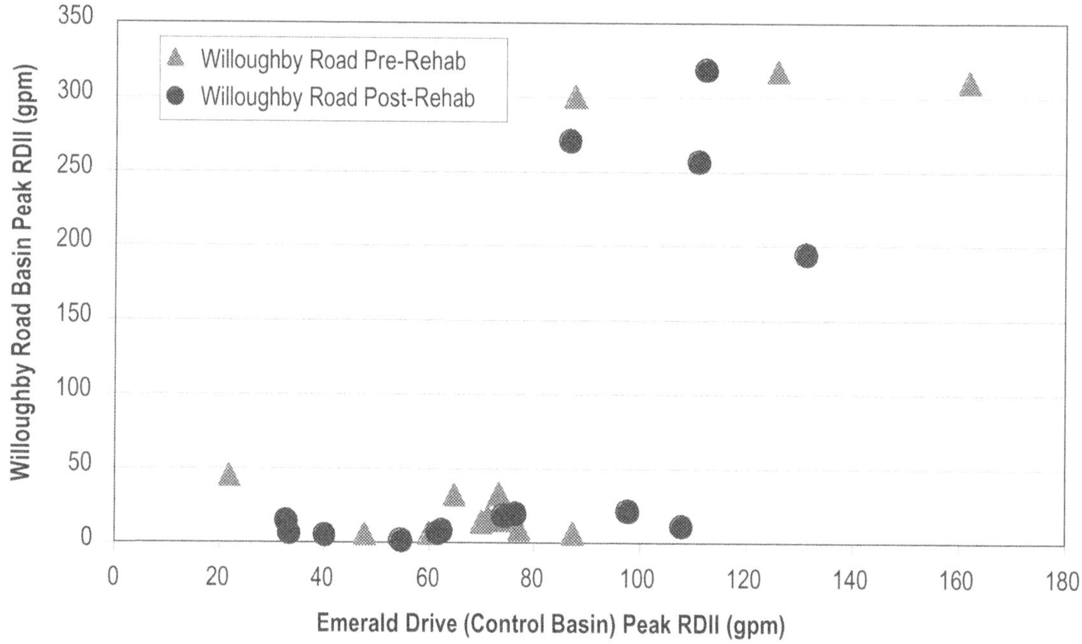

Figure 5-24. Comparison of Peak RDII for McCandless Moreland Road Basin and Control Meter.

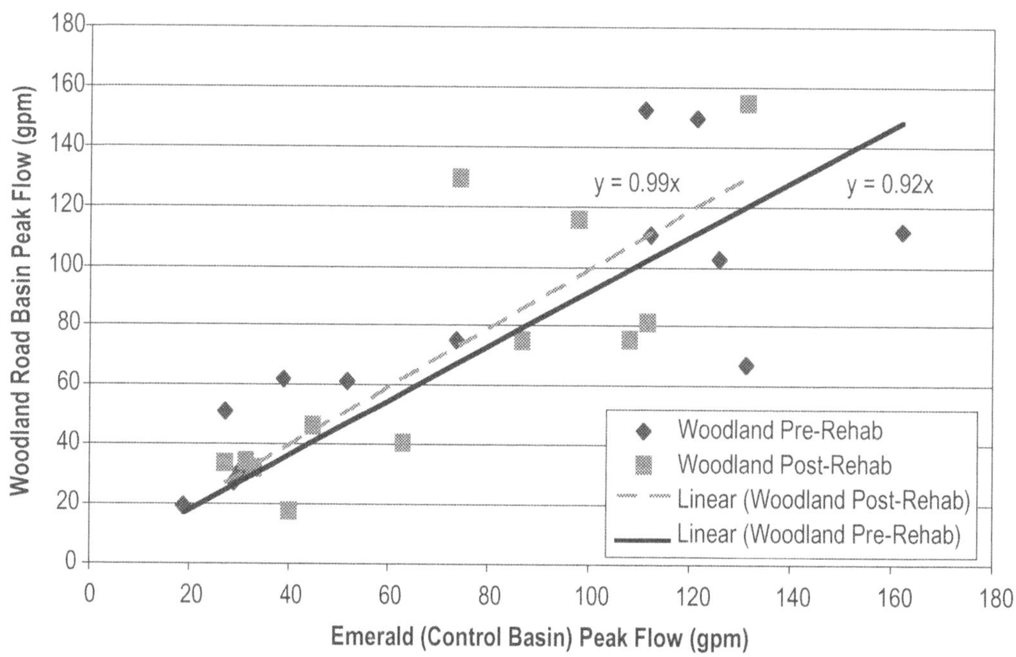

Figure 5-25. Comparison of Peak RDII for McCandless Woodland Road Basin and Control Meter.

Willoughby Road Basin. Figure 5-23 shows a comparison of peak RDII for the Willoughby Road and control basins; however, the data does not lend itself well to a best-fit line. The Willoughby Road Basin (Sloop Meter) was found to exhibit extraordinary peak RDII when daily precipitation exceeds about 1.0–1.5″. These large peaks are evident in Figure E-1 in Appendix E. This observation leads to the possibility that there could be an indirect sewer system connection to storm drainage, or that the groundwater may get high enough to enter the sewer system by defects in the upper laterals.

Moreland Road Basin. Figure 5-24 shows comparison data for the Moreland Road Basin. The plots of the Moreland Road and Woodland Road systems against the corresponding peaks in the control basin (Emerald Meter) indicate no detectable change in the ratio.

Calibrations for the Moreland Road Basin (Inwood Meter) are shown in Figures E-1 through E-4 in Appendix E. These figures provide RDII data for the Moreland Road system compared with the control basin (Emerald Meter). The figures show that peaks in the Moreland Road system were sometimes higher and sometimes lower than Emerald Meter in the pre-rehabilitation period and in the post-rehabilitation periods after main line rehabilitation and lateral replacement. The regression of peak on peak, however, indicates that the Moreland Road peaks were about 120 percent of the control basin peaks on average.

Woodland Road Basin. Figure 5-25 shows comparison data for the Woodland Road Basin. The plots of the Moreland Road and Woodland Road systems against the corresponding peaks in the control basin (Emerald Meter) indicate no detectable change in the ratio.

Authority staff performed observations of flow conditions in the laterals after construction of the lateral rehabilitations. Using the newly installed observation ports, the staff inspected the ports during periods of varying rainfall intensity and during dry periods after rainfall events.

Table 5-37 presents precipitation conditions when excessive private lateral flows were observed.

Table 5-37. Excessive Lateral Flow Observations.

Date	*Precipitation Conditions*
February 13, 2001	Wet—0.72″ on that day
February 26, 2001	Dry for preceding 7 days
March 7, 2001	Mostly dry for preceding 7 days
March 12–14, 2001	Little rain—wet for preceding week
March 21, 2001	Wet antecedent period—0.35″ on 21st; 0.26″ on preceding day; wet for preceding 2 weeks
April 16, 2001	Mostly dry for preceding 7 days

The lateral observations indicated that 54% of the laterals in the Moreland Road and Woodland Road sub-basins exhibited steady clear flow during the observations. The steady clear flows are reflective of groundwater that is continuing to infiltrate into the sewer system from defects located on private property. Only one of 17 laterals in the Willoughby Road sub-basin exhibited clear steady flow during similar observations.

5.5.6 Conclusions

Analysis of the flow data from the rehabilitated basins compared to the control basin indicates no detectable reduction in peak WW-related flows. This is consistent with the sewer lateral observations made by the McCandless Township Sanitary Authority using the newly installed sewer lateral inspection ports. Those observations indicated that 54% of the sewer laterals in the Moreland Road Basin and Woodland Road Basin exhibited steady clear flow indicative of groundwater during three observations.

It is possible that repairing the main line sewers may result in groundwater in the trenches rising to the point that it enters the sewer through defects in the upper sewer laterals.

5.6 Milwaukee Case Study

As an element of an RDII reduction demonstration project for the MMSD in Wisconsin, this case study analysis evaluated the effectiveness of manhole rehabilitation for a basin within the city of Milwaukee.

5.6.1 Project Background

The initial analysis of the area occurred in 1996, before manhole rehabilitation. In 2002, an additional analysis was completed that documented and assessed the impacts of manhole rehabilitation that had occurred between 1996 and 2000.

Figure 5-26 shows the manhole rehabilitation study area, which was evaluated before and after manhole rehabilitation.

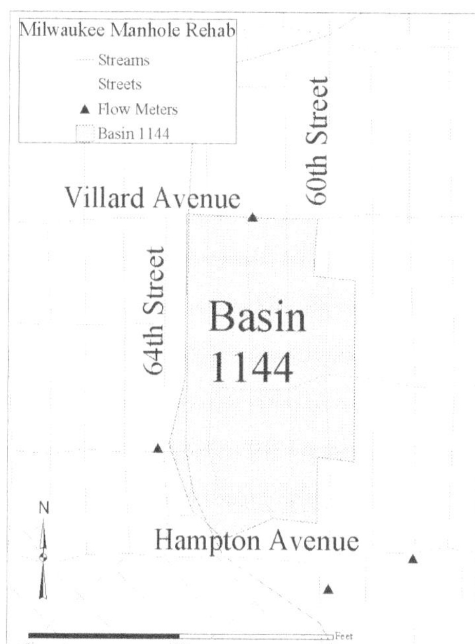

Figure 5-26. Milwaukee Basin 1144 Study Area.

The topography of Basin 1144 is predominantly flat, with soils that have an Ozaukee-Morley-Mequon association. These soils are well drained to somewhat poorly drained and have a subsoil of silty clay loam and silty clay, formed in thin loess and silty clay loam glacial till, on moraines (*Soil Survey of Waukesha and Milwaukee Counties Wisconsin*, USDA, 1998). Basin 1144 is primarily single-family residential, although a church is located in the southeast corner of the basin.

Table 5-38 summarizes the characteristics of the Basin 1144 study basin.

Table 5-38. Characteristics of Milwaukee Basin 1144 Study Basin

Basin	*Area (acres)*	*Sewer System (feet)*				*House Laterals*	*Sanitary Sewers (idm)*	*Number of Manholes*
		8″ Diameter	*10″ Diameter*	*15″ Diameter*	*18″ Diameter*			
1144	62	8,926′	0′	341′	18′	309	14.55	44

Basin 1144 totals 62 acres and is bounded by North 64th Street and North 60th Street to the west and east, and Custer Avenue and West Hampton Street to the north and south. The entire basin connects to separate storm and sanitary sewers. Basin 1144 discharges into a 72″ metropolitan interceptor system sewer on West Fairmount Avenue, west of North 64th Street. Basin 1144 contained 8,315 LF of separated sanitary sewer pipe in 1996 (pre-construction period) and 9,284 LF in 2,000 (post-construction period). In 2000, the sanitary sewer pipe within Basin 1144 was 8, 15, and 18″ in diameter, for a total of 14.55 idms. Manholes in the basin number 44.

On July 14, 1953, the city of Milwaukee adopted Ordinance Number 302, File Number 52-2153-d, effective January 1, 1954, that required footing drains on new buildings to be connected to a deep storm sewer if available. The assumption, therefore, is that sanitary sewer connections built in 1954 or later do not have foundation drains connected to the sanitary sewer. In 1996, Basin 1144 had 315 sewer connections, of which 290 were installed before 1955. The assumption, therefore, is that 92% of the connections in Basin 1144 have foundation drains connected to the sanitary sewer.

Basin 1144 has two pump bypasses. Bypass 0303 is located on West Fairmount Avenue, west of North 63rd Street, and was installed in 1973. Bypass 232 is located west of North 62nd Street and was installed in 1978. It is unknown what rehabilitation work was completed within this basin before 1996.

Peak flow problems simulated in Basin 1144 during the pre-construction period (before rehabilitation) included some surcharging on the 5-, 10-, 25-, and 50-year storm events. According to the *Sanitary Sewer Flow Monitoring and System Evaluation Report, Volume 1*, prepared by Rust Environment and Infrastructure in January 1997, the surcharging anticipated for the five- and 10-year design storms was not enough to cause basement backups; however, the surcharging during the 25- and 50-year events was significant enough to cause basement backups. A peaking factor for this basin was not identified in this report.

5.6.2 Monitoring Data

This section describes the flow and rainfall data collected as part of this sewer rehabilitation project. Monitoring of Basin 1144 comprised monitoring the flows and collecting the rainfall data before and after rehabilitation.

5.6.2.1 Flow Data

During the pre- and post-construction periods, Basin 1144 was monitored with Isco 4150 area-velocity flow meters. Velocity, wastewater level, and internally calculated flow were gathered during the monitoring process. Calibration of the flow monitors occurred weekly and consisted of checking the depths, changing the offset, and/or verifying the velocity. In addition, each meter was annually sent to the manufacturer for recalibration. Routine maintenance of the flow monitors occurred weekly, when the flow meter data was collected, and consisted of removing any objects caught on the meter probe. Cretex surcharge level indicators were installed at each flow meter site and were also visited weekly to note any surcharge and to reset the level indicator. Backups for the collected data were made with Isco's Flowlink version 3.2 software for the pre-rehabilitation data and version 4.1 for the post-rehabilitation data.

During the pre-construction period, the length of the flow-monitoring record for Basin 1144 was from March 3–November 11, 1996, and had a time step of 10 minutes. Post-construction flow monitoring occurred from May 23–August 23, 2000, and the data was stored at a time step of five minutes.

5.6.2.2 Rainfall Data

Rainfall information was obtained from the MMSD network and provided by MMSD in electronic format. Basin 1144 used rain gauges WS1207 and WS1202, with a weighted influence corresponding to each gauge's distance from the project sites. Gauge WS1207—located at 8414 West Florist Avenue, Milwaukee, Wisconsin—has a time step of 1 hour and a length of record from January 1–December 30, 1996 (pre-construction period) and January 1–December 31, 2000 (post-construction period). Gauge WS1202—located at 5335 North Teutonia, Milwaukee, Wisconsin—has a time step of one hour and a length of record from January 1–December 30, 1996 (pre-construction period) and January 1–December 31, 2000 (post-construction period). Long-term RDII simulations used rainfall data acquired from NOAA from 1940 through 2001 for Mitchell Field.

5.6.3 SSES Information

Activities performed during the SSES investigations for the study areas included aboveground manhole inspections, smoke testing, sewer televising, building inspections, and dyed water flooding of suspected sources identified during smoke testing. The results of these investigations are unknown.

The Sanitary Sewer Flow Monitoring and System Evaluation Report, Volume 1, completed in January 1997 by Rust Environment and Infrastructure, recommended that basins with detected infiltration that exceeded 20,000 gpd/in-mile have manhole inspection and internal television inspection during high groundwater conditions to identify infiltration sources into the sanitary sewer system. Basin 1144 had calculated infiltration amounts of 69,893 gpd/in-mile (Rust, 1997).

5.6.4 Project Details

This section provides details on the work performed during this manhole rehabilitation project, including a discussion of project costs for Milwaukee Basin 1144.

5.6.4.1 Construction Details

Table 5-39 describes the rehabilitation completed and quantity of manholes that were repaired. These construction and rehabilitation methods were selected because ground manhole inspections found non-watertight covers or covers with pick/vent holes, frame/chimney interfaces that were not sealed, and misaligned frames. "Defects which exist in the upper portion of a manhole such as the chimney, frame, and cover usually are considered as potential sources of inflow. Defects in the lower sections (cone, barrel, bench, etc.) are usually sources of infiltration or groundwater intrusion" (Rust, 1997).

Table 5-39. Milwaukee Basin 1144 Rehabilitation Performed.

Description of Rehabilitation	*Units*	*Quantity*
Install Solid Gasketed Lids	Each	37
Plug Lids	Each	3
Repair Internal Seals	Each	12
Vertical Chimney Brickwork	V.F.	1.2
Vertical Corbel Brickwork	V.F.	1.8

5.6.4.2 Construction Costs

Basin 1144 (ENR 6342) cost approximately $10,836 (2002 dollars).

5.6.5 Post-Construction Data Evaluation

This section discusses the analysis performed before and after construction conditions. The discussion includes a review of data quality issues and an overall summary of RDII reduction success. The following paragraphs discuss the calibration and analysis results.

5.6.5.1 Calibration

Figures F-1 through F-4 in Appendix F represent calibration plots comparing the model simulations with measured RDII flow and rainfall. Both pre-rehabilitation monitoring periods contained significant peaks against which the models could be calibrated. The post-rehabilitation period was more difficult to calibrate because there were many small rainfall events and no exceptionally large rainfall events with which to calibrate.

5.6.5.2 Results

Tables 5-40 through 5-42—and Figures 5-27 through 5-29—summarize Basin 1144 long-term simulation results. The percent removal effectiveness for Basin 1144 is higher than anticipated. The model results are supported by comparison of peak measured RDII to antecedent rainfall (see Figure F-5 in Appendix F).

Table 5-40. Pre- and Post-Construction Peak-Hour RDII Conditions for Milwaukee Basin 1144.

Condition	*5-Year Peak-Hour RDII*			*10-Year Peak-Hour RDII*			*20-Year Peak-Hour RDII*		
	mgd	*gpad*	*gpd/idm*	*mgd*	*gpad*	*gpd/idm*	*mgd*	*gpad*	*gpd/idm*
Pre-Construction	3.732	60,190	256,481	4.375	70,562	300,675	4.944	79,741	339,790
Post-Construction	2.043	32,947	140,393	2.432	39,225	167,143	2.776	44,773	190,784
RDII Removal Effectiveness	1.689	27,243	116,088	1.943	31,337	133,532	2.168	34,968	149,006
Percent Removal Effectiveness	45%			44%			44%		

Table 5-41. Pre- and Post-Construction Peak-Day RDII Conditions for Milwaukee Basin 1144.

Condition	*5-Year Peak-Day RDII*			*10-Year Peak-Day RDII*			*20-Year Peak-Day RDII*		
	mgd	*gpad*	*gpd/idm*	*mgd*	*gpad*	*gpd/idm*	*mgd*	*gpad*	*gpd/idm*
Pre-Construction	1.396	22,521	95,967	1.710	27,582	117,533	2.013	32,461	138,322
Post-Construction	0.840	13,554	57,756	1.045	16,856	71,824	1.237	19,960	85,051
RDII Removal Effectiveness	0.556	8,967	38,211	0.665	10,727	45,709	0.775	12,502	53,271
Percent Removal Effectiveness	40%			39%			39%		

Table 5-42. Pre- and Post-Construction Peak-Month RDII Conditions for Milwaukee Basin 1144.

Condition	*5-Year Peak-Month RDII*			*10-Year Peak-Month RDII*			*20-Year Peak-Month RDII*		
	mgd	*gpad*	*gpd/idm*	*mgd*	*gpad*	*gpd/idm*	*mgd*	*gpad*	*gpd/idm*
Pre-Construction	0.166	2,676	11,403	0.186	3,006	12,811	0.203	3,277	13,966
Post-Construction	0.079	1,274	5,430	0.091	1,465	6,241	0.100	1,607	6,850
RDII Removal Effectiveness	0.087	1,402	5,974	0.096	1,542	6,570	0.104	1,670	7,116
Percent Removal Effectiveness	52%			51%			51%		

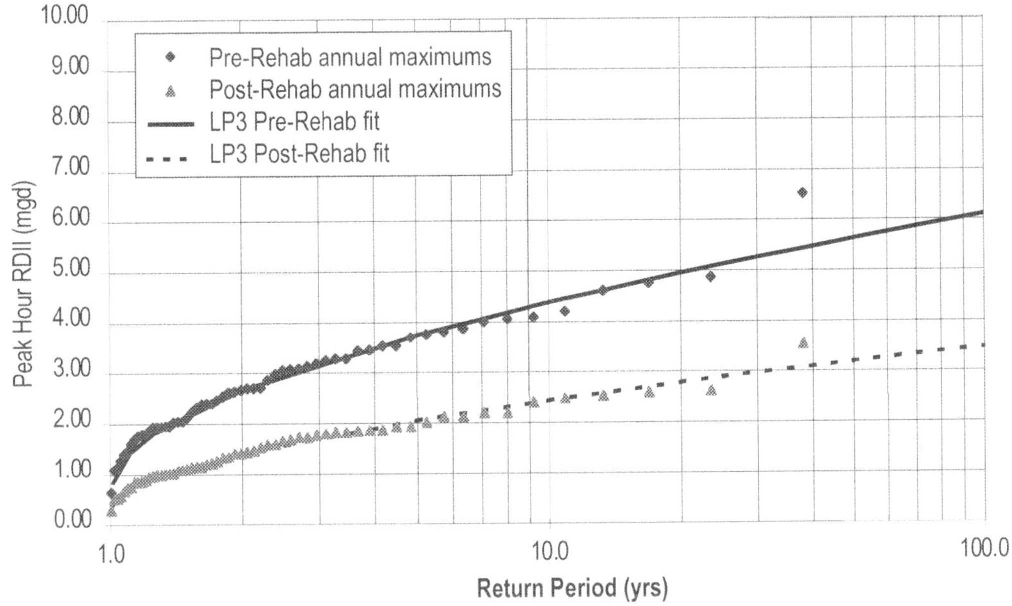

Figure 5-27. Return Period for Peak-Hour RDII for Milwaukee Basin 1144.

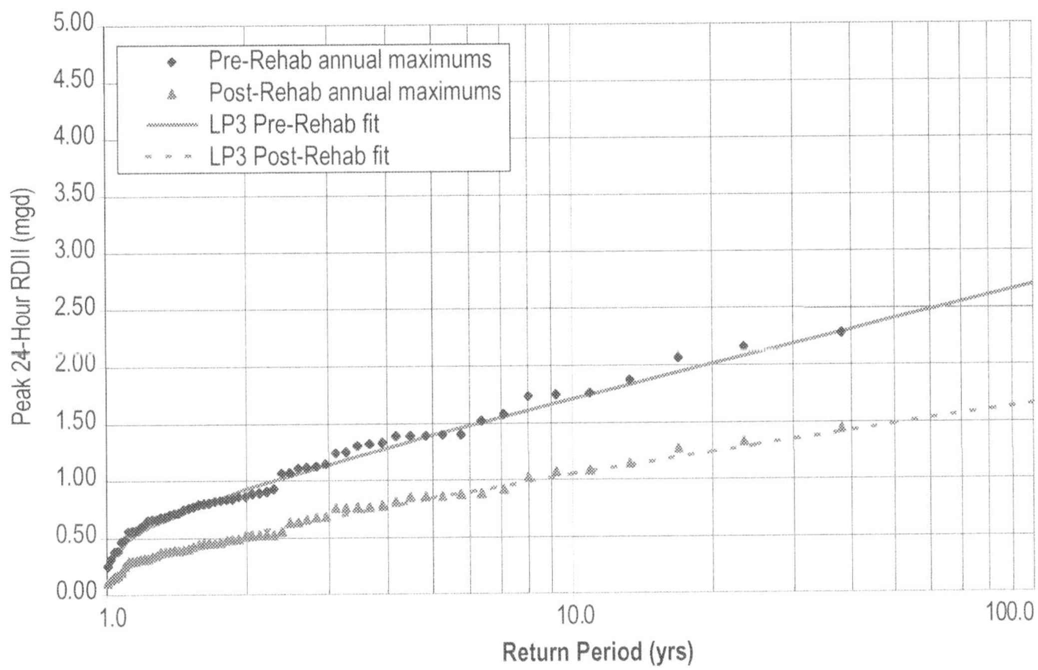

Figure 5-28. Return Period for Peak-Day (24-Hour Average) RDII for Milwaukee Basin 1144.

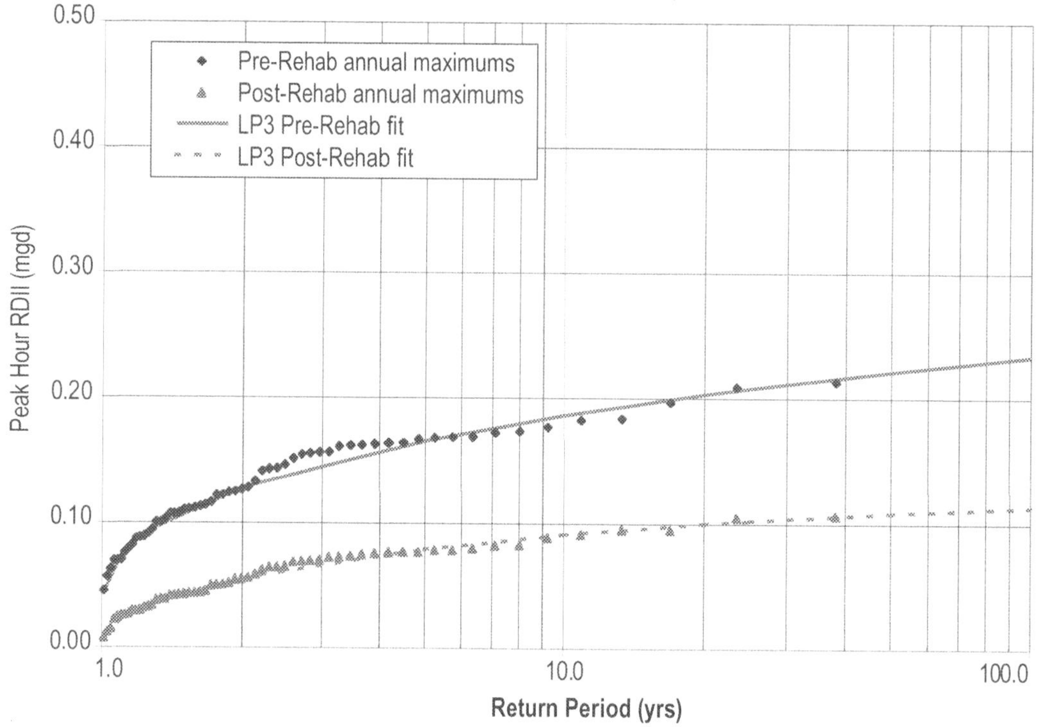

Figure 5-29. Return Period for Peak-Month (30-Day Average) RDII for Milwaukee Basin 1144.

5.6.5.3 Cost-Effectiveness Analysis

Effectiveness of the manhole rehabilitation projects was analyzed against the cost of the rehabilitation for Basin 1144.

Table 5-43 summarizes the cost-effectiveness results for Basin 1144.

Table 5-43. Peak RDII Removal Cost-Effectiveness Measures for Milwaukee Basin 1144.

RDII Condition	*RDII Removal Cost-Effectiveness, \$/gpd*[a,b]		
	Peak Hour	*Peak Day*	*Peak Month*
5-year	\$0.01	\$0.02	\$0.12
10-year	\$0.01	\$0.02	\$0.11
20-year	\$0.00	\$0.01	\$0.10

[a]Construction costs only.
[b]Costs are adjusted to 2002 dollars.

5.6.6 Conclusions

In conclusion, if the manhole rehabilitation is constructed properly and if manhole lids are a major inflow source, manhole rehabilitation may achieve 40% reduction as seen with Basin 1144.

5.7 Summary and Conclusions Concerning Case Studies

The six case studies present detailed information and analysis concerning 12 RDII reduction projects constructed in the United States.

Table 5-44 presents the rehabilitation techniques used, together with the estimated five-year peak RDII flows before and after rehabilitation and the corresponding reductions.

Table 5-44. Summary of Rehabilitation Techniques and RDII Removal for Each Project

Project	*Type of Project*					*5-Year Peak-Hour RDII (gpad)*		
	Manhole Rehabilitation	*House Lateral Rehabilitation*	*Sewer Main Rehabilitation*	*Roof Drain Disconnection*	*Foundation Drain Disconnection*	*Pre-Construction*	*Post-Construction*	*Percent Reduction*
Oak Creek					✓	5,474	2,851	48%
Honolulu—Basin A	✓		✓			8,063	7,666	5%
Honolulu—Basin B	✓		✓			12,468	12468	nil
LOTT—OL22 Phase I		✓ (lower)	✓			19,047	15,807	17%
LOTT—OL22 Phase II		✓ (upper)	✓	—[a]	—[a]	15,807	7,421	53%
LOTT—OL24		✓				1,219	573	53%
King County—Kent	✓	✓	✓			23,680	5,300	78%
King County—Issaquah	✓	✓	✓			6,877	3,390	51%
McCandless—Moreland Road Basin		✓		✓		—	—	nil
McCandless—Woodland Road Basin	✓	✓		✓		—	—	nil
McCandless—Willoughby Road Basin		✓	✓				—	nil
Milwaukee Basin 1144	✓					60,190	27,243	45%

[a]Incidental disconnection of downspouts and foundation drains occurred.

The purpose of analyzing these projects with a common approach was to be able to compare them in an appropriate manner. Many previous attempts at comparing RDII projects compared reported RDII reductions, without any verification as to the accuracy of the figures, attention to the differences in methods of estimating reductions, or attention to the differences in project circumstances. This section carries the analyses beyond the reductions achieved to examine these issues. The comparisons provided here are:

- Pre-construction RDII levels
- Residual RDII after project completion
- Cost-effectiveness of RDII removal
- Effectiveness of alternative rehabilitation techniques

5.7.1 Comparison of Pre-Construction and Residual RDII Levels

The case study projects represent a variety of RDII conditions before construction. For the most part, all of the projects were chosen by the implementing agencies because they were seen as opportunities to reduce WWFs in "problem basins." Each project featured in this report, therefore, could be considered to have a large amount of RDII. For proper comparisons of RDII levels, this report presents the pre-construction RDII, which is normalized by the number of service connections upstream of the flow monitor. Ultimately, the success of an RDII removal project is judged by its ability to reduce WWFs at some location downstream.

Figure 5-30 shows pre-construction and residual RDII levels for the projects in terms of gpd per service connection. It is interesting to note that the residual RDII in some basins is actually higher than the pre-construction RDII in other basins. Such comparisons lend credibility to the argument for site-specific RDII removal data as a basis for determining a direction for percent RDII control.

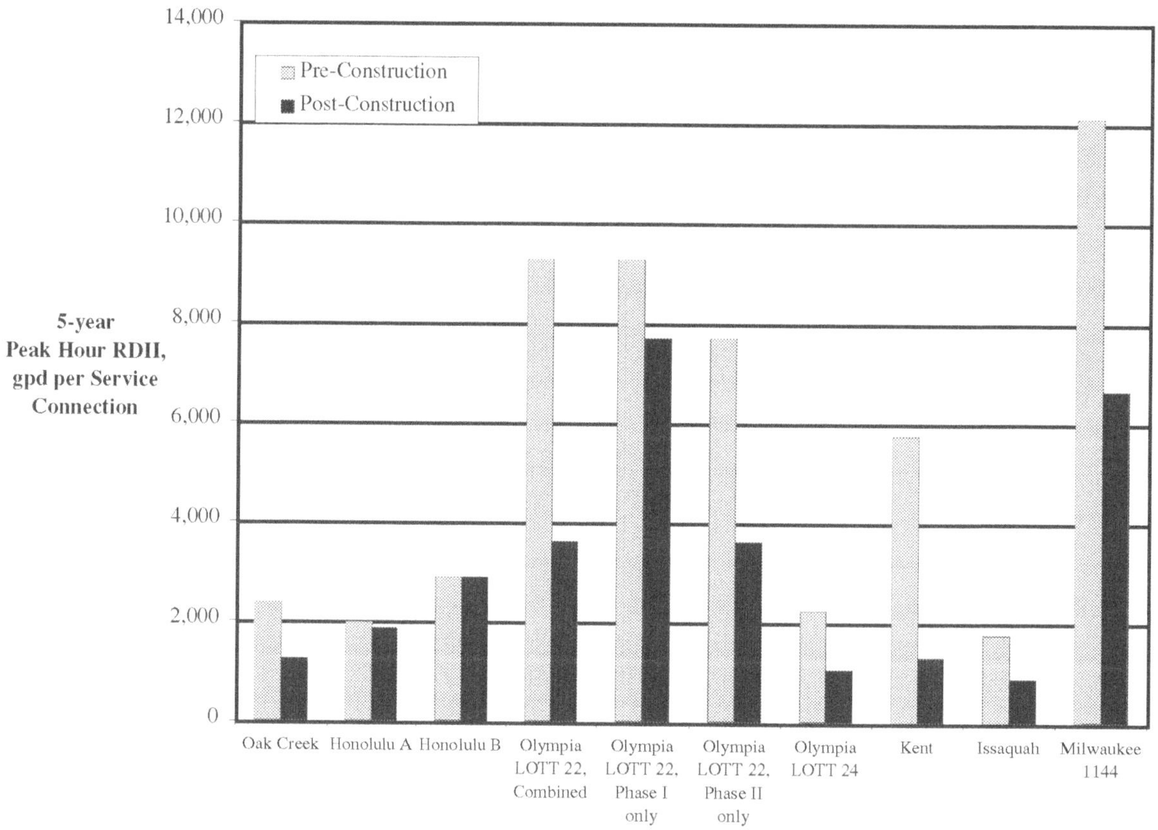

Figure 5-30. Comparison of Pre-Construction and Residual RDII Levels.

Figure 5-31 shows the percent RDII removal of each project.

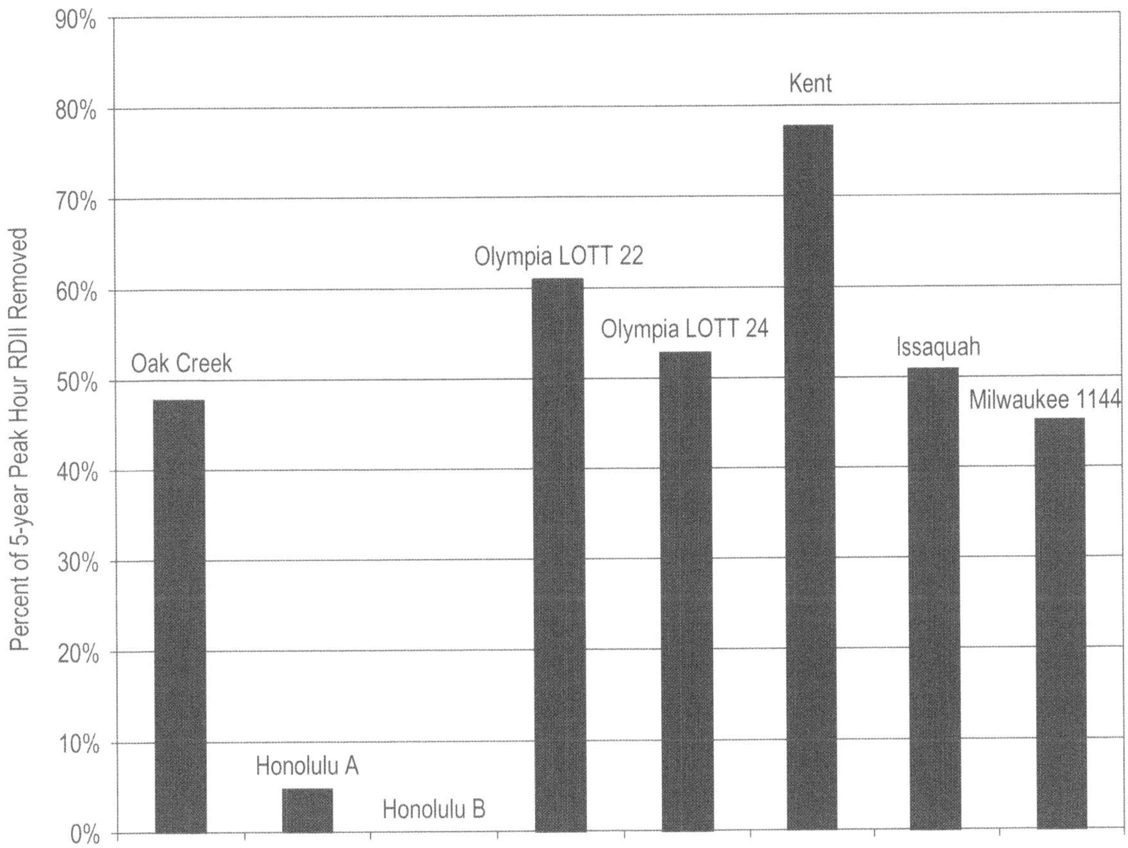

Figure 5-31. Comparison of RDII Removal Achieved by Case Study Projects.

5.7.2 Comparison of RDII Removal Cost-Effectiveness

The concept of "cost-effective" RDII removal traditionally has consisted of either comparing costs to the cost of transport and treating the RDII or finding the point at which RDII removal costs are higher than the return on investment. This point is commonly referred to as the "knee of the curve."

Figure 5-32 shows this cost-effective definition. Each curve on the plot represents a possible relationship between in-system RDII before construction and the cost per RDII flow rate removed. The knee of the curve for each represents the point at which the cost to remove RDII accelerates to a point of diminishing return. Although these curves are theoretical, they make sense to many utility managers. Unfortunately, these curves have not been compared appropriately with field data.

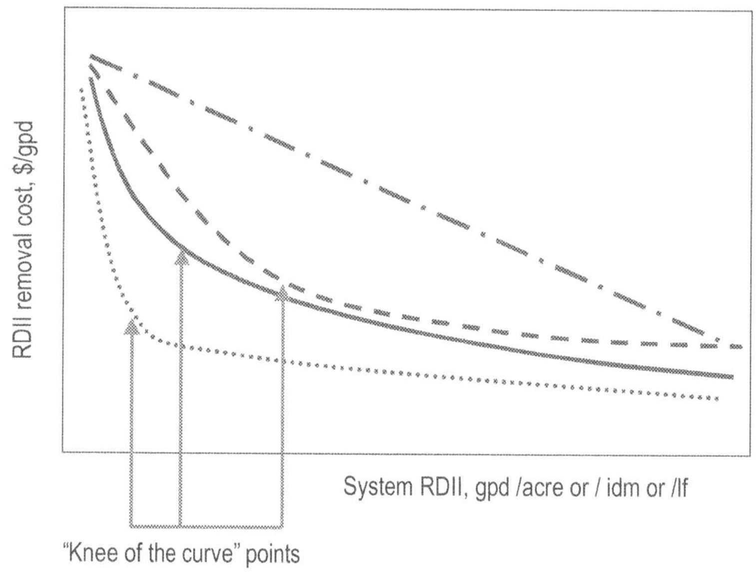

Figure 5-32. Theoretical Cost-Effectiveness Curves for RDII Removal.

The case study analyses were evaluated in such a manner that makes comparisons possible, including a cost-effectiveness determination.

Figures 5-33 and 5-34 show a plot of pre-construction RDII levels against the cost-effectiveness of removal for select projects. Neither Honolulu project was included on this plot. These two projects, by inspection, are seen as not being cost effective given the cost and poor RDII removal rates.

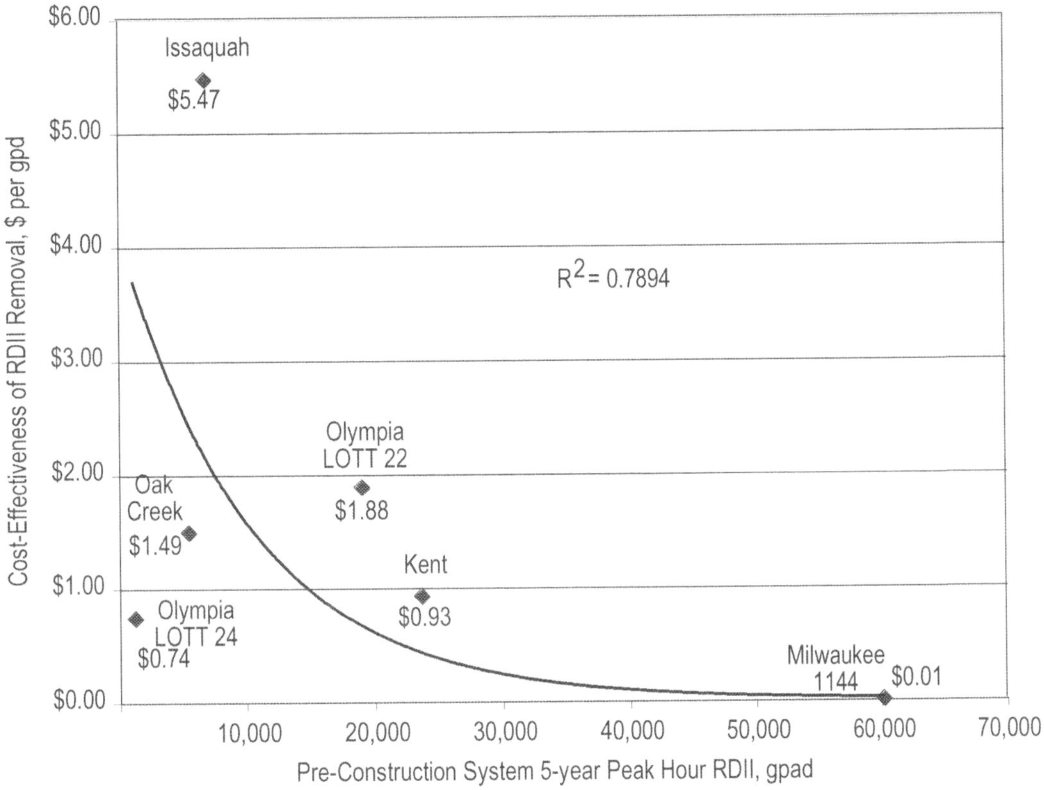

Figure 5-33. Cost-Effectiveness Comparison Curve (gpad) for Select RDII Case Study Projects.

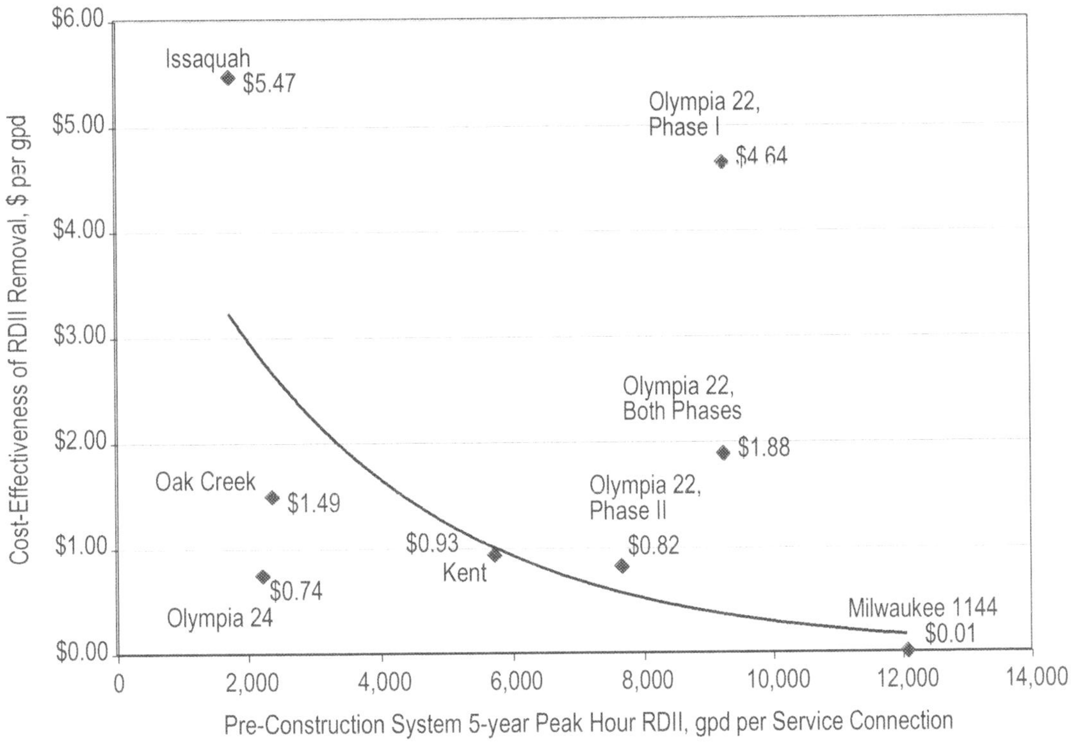

Figure 5-34. Cost-Effectiveness Comparison Curve (gpad per Service Connection) for Select RDII Case Study Projects.

As for the projects that were included in the graph, a curve can be fit to the data. Although one can discern a knee of the curve from the fit (at approximately 14,000 gpad for a five-year peak hour RDII flow), it is important to note that this curve is based on successful RDII projects and not all RDII projects. An important additional step in determining cost-effectiveness is to compare these figures with local costs to transport, store, and treat the peak RDII flow rate and volume, which can be referred to as "facilities costs." If current facilities can handle the system RDII without violating any level of service objectives for the system, the facilities costs to compare with RDII removal costs would be limited to typical O&M costs (e.g., power, chemicals, personnel). If current facilities are incapable of achieving the utility's desired level of service, RDII removal costs must be compared with facilities costs that represent the nominal cost for capacity enhancements that would achieve the level of service, in addition to the aforementioned O&M costs.

5.7.3 Removal Effectiveness of Alternative Rehabilitation Techniques

The case studies present a range of approaches to rehabilitation that reduces RDII. These include simple manhole rehabilitation; repair or replacement of the utilities' sewers in the PROW with or without repair of the service lateral to the property line; repair or replacement of just the private sewers; and repair or replacement of all sewers, both public and private. There is also a range of approaches in addressing private sources, including foundation drains/sump pumps and roof drainage.

The results of the analyses of these projects, supported by the literature survey, strongly indicate that ignoring the private sewers puts utilities at risk of not reducing peak RDII flows to any significant degree. The projects that addressed sewers only in the PROW—even where roof drainage was addressed—achieved peak RDII removals from zero to a maximum of 17%, with the majority near the zero end of the scale. Projects that addressed the private sources only achieved peak RDII removals of 50% or more. From these observations, the conclusion is that the most effective approach to peak RDII removal begins with private sources. The first step in a program should address foundation drains connected to the sanitary sewer and roof drainage. Anecdotal evidence from the city of Salem, Oregon, indicates that a sump pump delivered up to 17,000 gpd to the public sewer (Ken Roley, city of Salem, personnel communication: flows were measured by a water meter attached to the pump discharge). Connected roof drains should, of course, be disconnected if found; however, the drainage should also be directed far enough away from the building envelope to prevent the drainage from percolating directly into the building sewer.

It is also necessary to address the issue of the migration of groundwater. Because of the entry of groundwater through other defects, removal of sump pumps and roof drainage where found may not achieve as significant peak RDII reduction as expected in some areas. Rehabilitating private sewers that are observed to be contributing RDII during CCTV inspections may result in the migration of water to other laterals or defects that were not previously leaking. This argues for comprehensive rehabilitation or at least a comprehensive survey of defects.

CHAPTER 6.0

Peak RDII Removal Projects—Reporting Protocol

For RDII removal project reports to benefit those outside the community where the work was performed, a standard reporting protocol should be followed. By using a standard reporting protocol, complete documentation of the project would be available. Such documentation would serve as a valuable resource to communities that are researching the best method of RDII evaluation/reduction. Previously performed projects could be evaluated and used to determine what would work effectively for a particular community.

Based on the research conducted on RDII removal projects across the United States, it is clear that an effectiveness evaluation reporting protocol for such projects does not currently exist. It would, however, be appropriate. The recommended overall elements of the reporting protocol include:

- Project background information
  - Flow-monitoring basin characteristics
  - Other SSES information
- Project details
  - Part of system targeted
  - Extent of construction within basin
  - Design and construction details
- Post-construction evaluation
  - Monitoring before and after construction
  - Modeling approach used and results

Sections 6.1 through 6.3 detail the type of information that would benefit readers in better understanding the project, from the setting to the purpose to the results. The authors believe that if such a reporting protocol were followed, the usefulness of RDII removal project data would extend beyond individual communities or consulting firms.

6.1 Project Background Information

Project background information includes a description of the sewer system and its surrounding. Flow-monitoring basin characteristics are key to understanding the magnitude of RDII that a system is experiencing and helps in comparative analyses with other projects. Many times, it is helpful to normalize RDII reduction results by characteristics of the flow-monitoring basin. The types of information that should be included as background are:

- Type of system: sanitary sewer, combined sewer, previously combined sewer, etc.
- Amount of sewer system: length, idms, number of manholes, diameters, sizes of lateral pipes
- Characteristics of sewer system: system age, past rehabilitation work, construction materials
- Size of basin: total area, sewered area, number of connections
- Natural setting: topography, native soil conditions, current land uses

Other SSES information will also be needed to expose the potential causes of RDII and help lead readers to reasons for selecting the construction methods. Types of SSES data that should be included in the document are:

- Sewer televising, smoke and dye testing, manhole and building inspections, etc.
- Other system information
- Existing sewer/manhole construction
- Adequacy of storm drainage system

6.2 Project Details

Project details should answer the questions "what was done and where?" This information should contain as much quantity data as possible to advance the knowledge base of RDII reduction construction techniques. Specific information to include is:

- What construction methods were selected and why
- What portion of flow-monitored basin was repaired
- Number of connections
- Length of sewer
- Number of manholes
- How the design was packaged
- How construction was procured
- How much the project cost, broken down into project elements
- Inflow sources
- Sewer main
- Manholes
- Private service laterals

In documenting the details of construction, it is as important to discuss what was not done as what was done. This is particularly true if construction did not affect a portion of the flow-monitoring basin. Many considerations can contribute to this happening, including lack of budget, different system conditions, different land use, and many other possibilities. Although construction documents are not typically bid to provide this data, careful preparation of bid documents can allow a utility to track better the costs of each type of RDII reduction.

Design packaging and construction procurement can be particularly challenging when private property facilities are included in the project. Because many utilities have not tried performing large sewer rehabilitation projects on private property, much remains to be learned from those that have.

6.3 Post-Construction Evaluation

Post-construction evaluation is the final important step in documenting the RDII project. This step comprises two key elements: (1 rainfall and flow monitoring and (2 data evaluation. Although many approaches are available to each, the choices made can influence the credibility of the conclusions. The flow-monitoring approach should be appropriate for the conditions of the sewer system. Proper site and flow-monitoring equipment selection should be performed.

Although monitoring data is the most important aspect of determining the success of an RDII project, it is typically the most weakly documented element of an RDII reduction report. The types of information that should be documented are:

- Type of monitor used
  - Depth
  - Velocity
  - Method of determining flow rate
- Frequency of calibration/maintenance, with logs of actions taken
  - Length of record
  - Rainfall gauge information, including type and distance relative to basin

The information must be gathered for both the pre- and post-construction period. Without this data, there will be some question as to the validity of the conclusions made on the project's success.

APPENDIX A

OAK CREEK CASE STUDY: SUPPLEMENTARY INFORMATION

Table A-1. Average Absolute Errors in Predicted vs. Measured Hourly Flows.

Pre-Rehabilitation		*Post-Rehabilitation*	
Dates	*% Error*	*Dates*	*% Error*
Aug. 28–Sept. 2, 1995	14	Aug. 5–6, 2000	31
Aug. 16–21, 1995	31	Aug. 17–18, 2000	30
Oct. 6–9, 1995	22	Sept. 11–12, 2000	21
June 17–20, 1996	11		

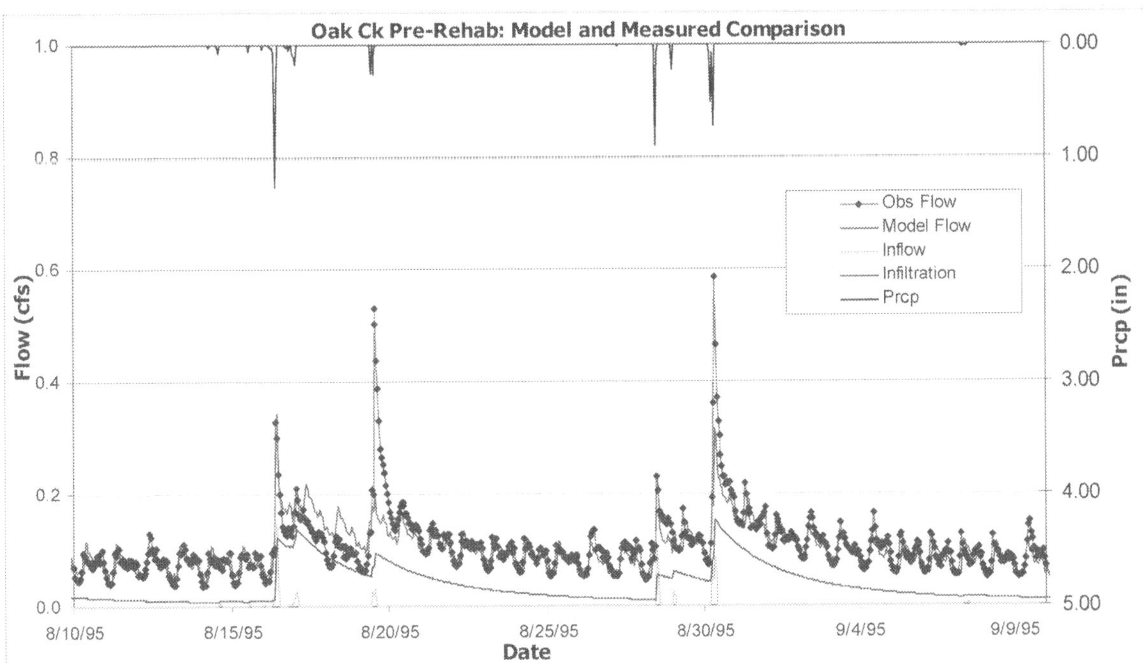

Figure A-1. Oak Creek Pre-Rehabilitation Calibration: August 10–September 9, 1995.

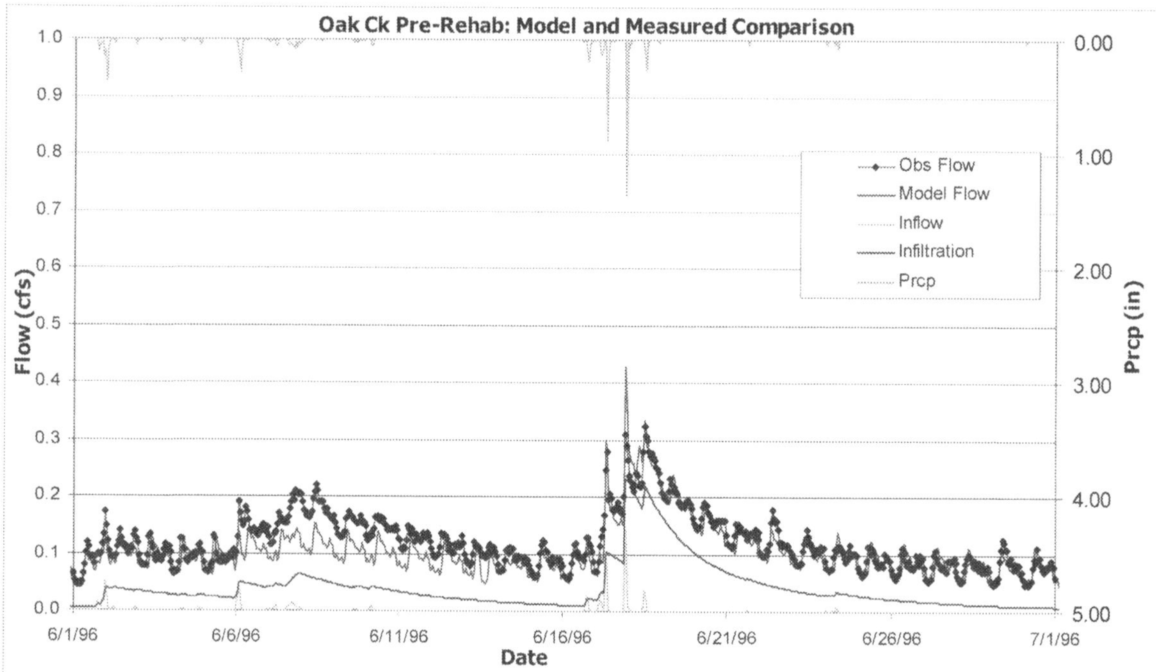

Figure A-2. Oak Creek Pre-Rehabilitation Calibration: June 1–July 1, 1996.

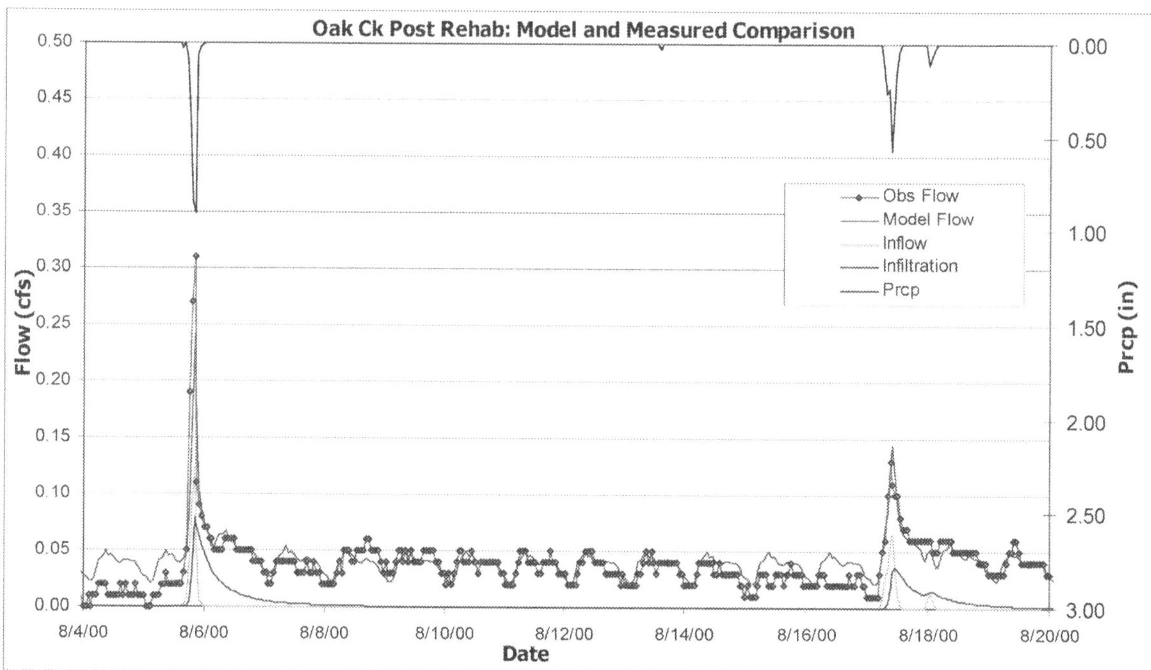

Figure A-3. Oak Creek Post-Rehabilitation Calibration: August 4–20, 2000.

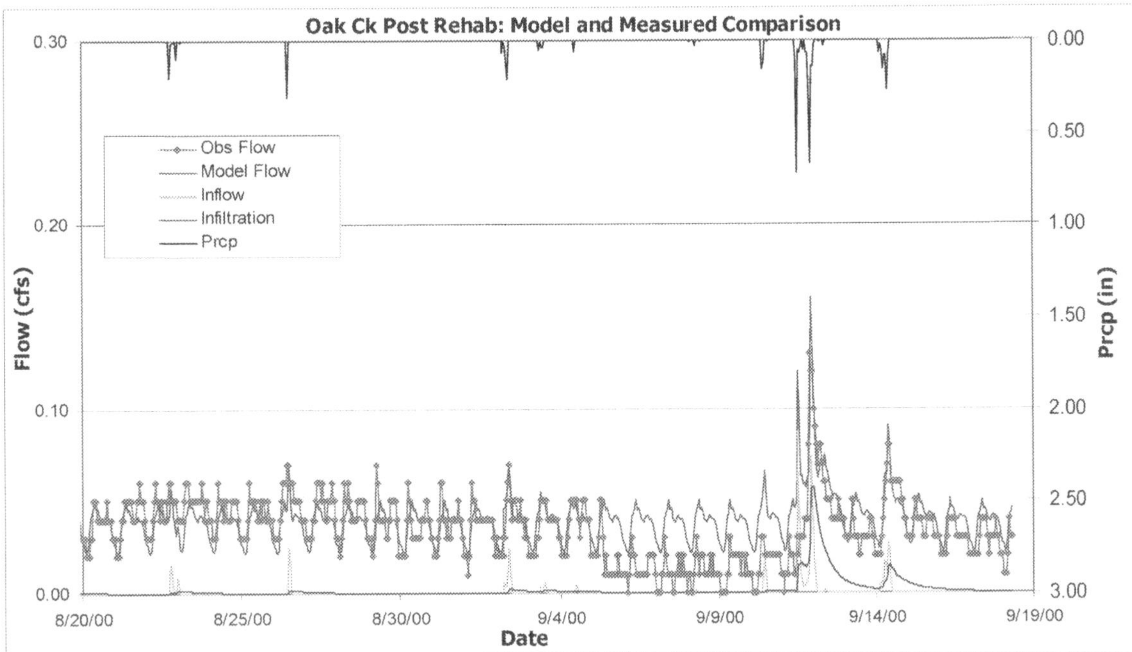

Figure A-4. Oak Creek Post-Rehabilitation Calibration: August 20–September 19, 2000.

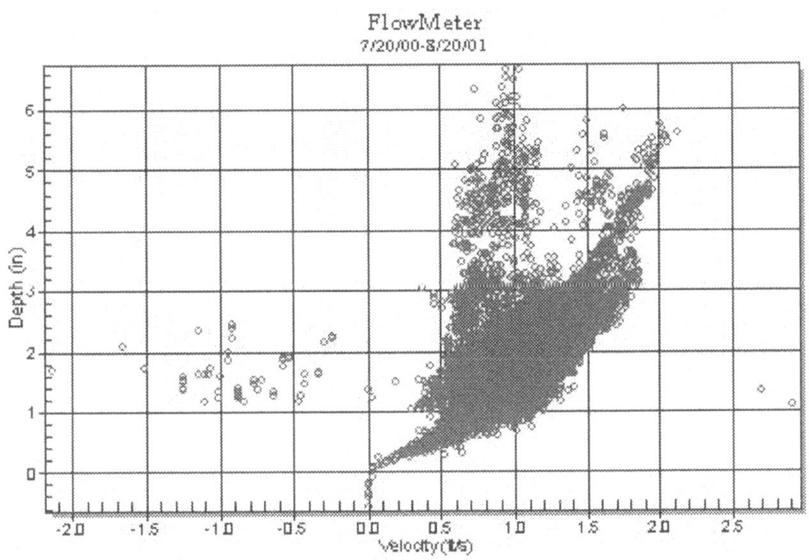

Figure A-5. Oak Creek Flow Meter OC0109 Post-Rehabilitation Velocity-Depth Scatter Plot of depth versus velocity.

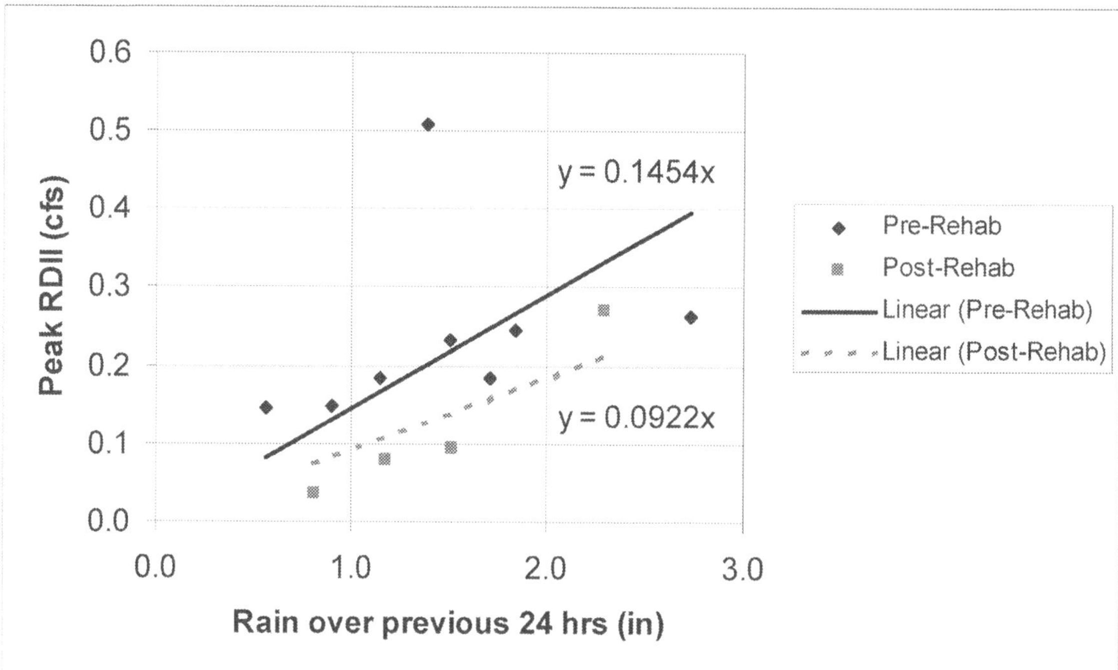

Figure A-6. Oak Creek Peak Q vs. RDII Plot (Peak Measured RDII vs. Sum of Rainfall Previous 24 Hours).

HONOLULU CASE STUDY: CALIBRATIONS AND VELOCITY-DEPTH SCATTER PLOTS

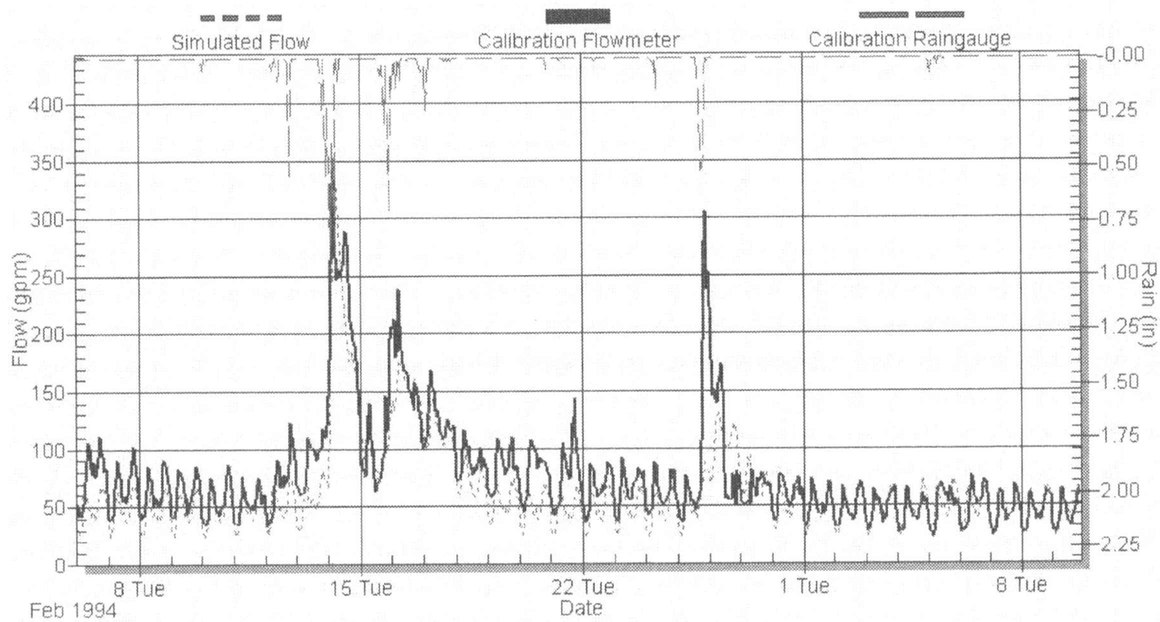

Figure B-1. Meter 2-203 (Basin A) Pre-Rehabilitation Period: February 1994.

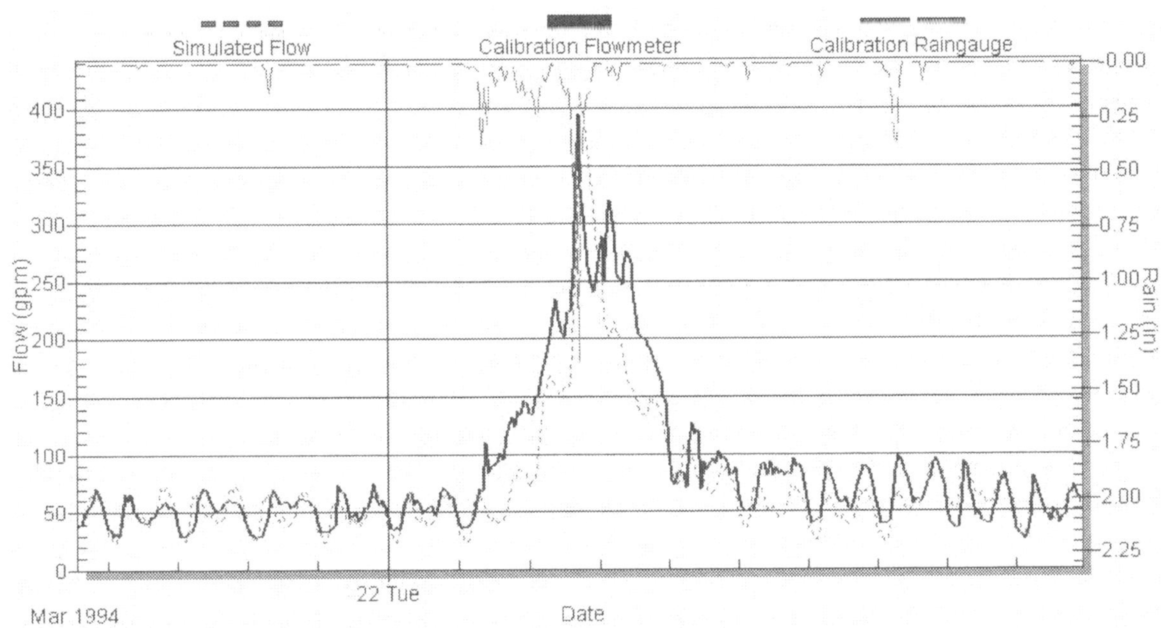

Figure B-2. Meter 2-203 (Basin A) Pre-Rehabilitation Period: March 1994.

(Record shows surcharge of up to 90 inches.)

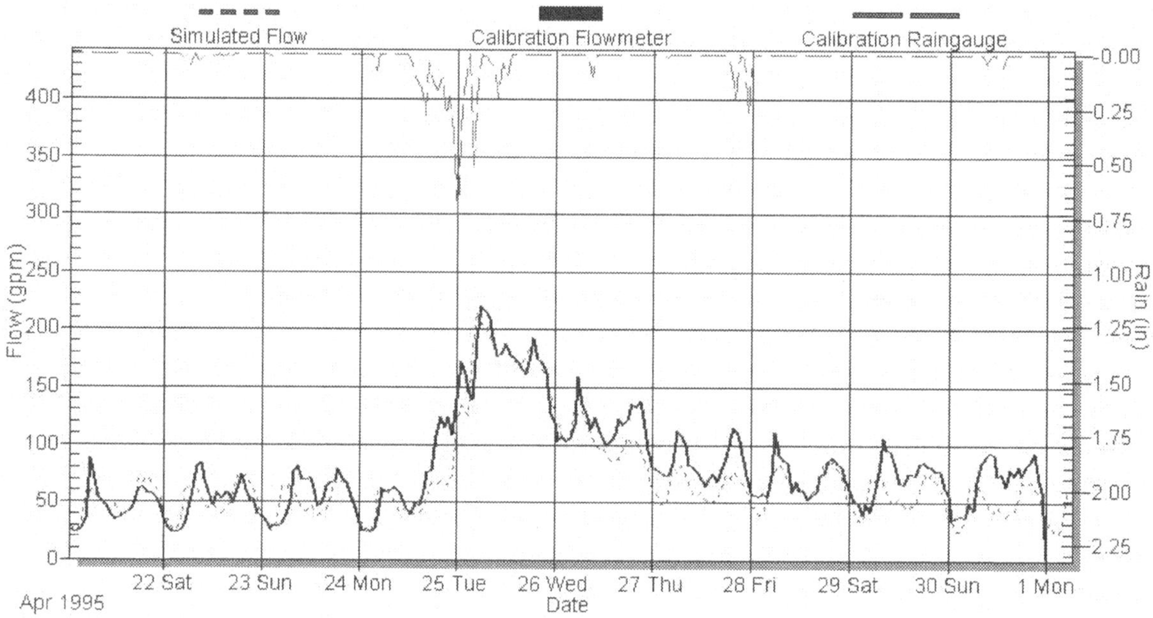

Figure B-3. Meter 2-203 (Basin A) Pre-Rehabilitation Period: April 1995.

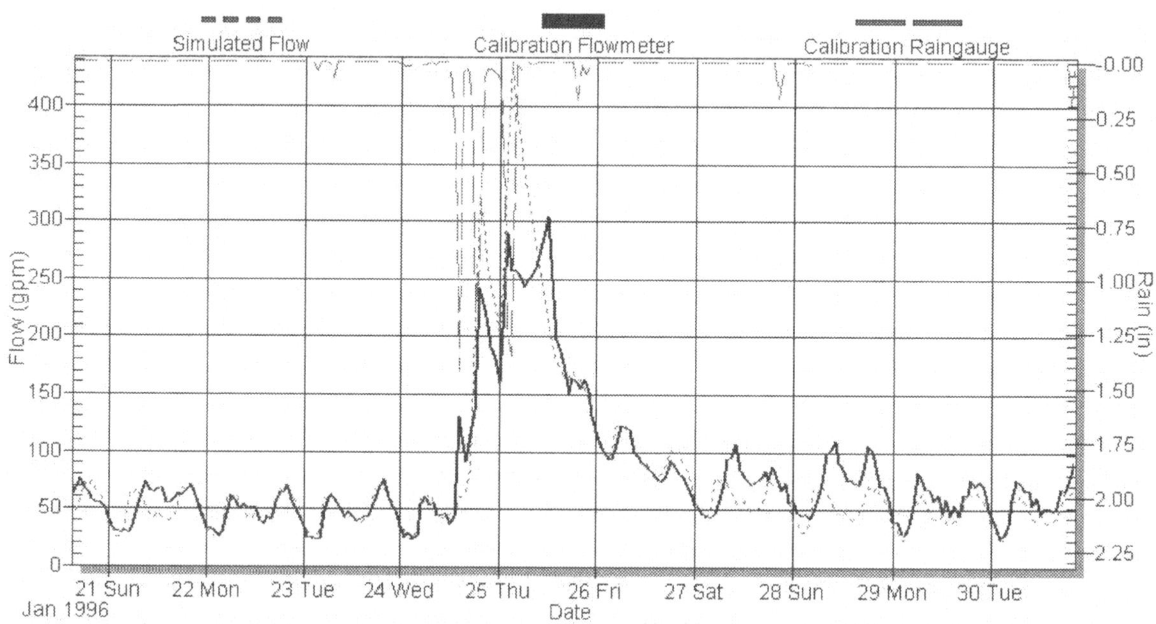

Figure B-4. Meter 2-203 (Basin A) Pre-Rehabilitation Period: January 1996.

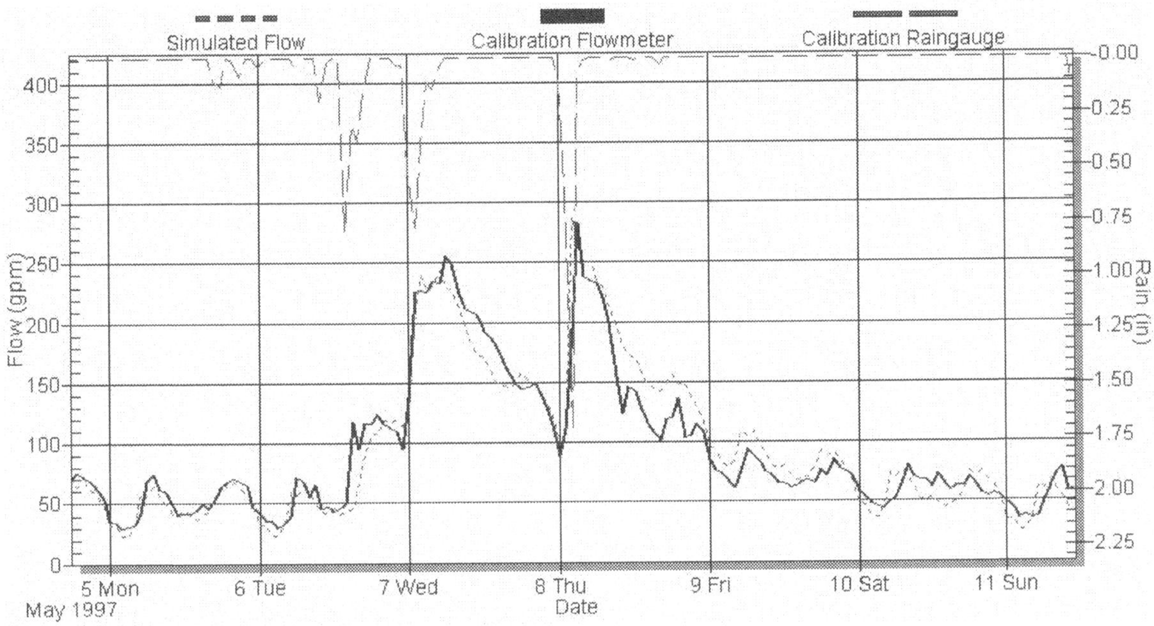

Figure B-5. Meter 2-203 (Basin A) Post-Rehabilitation Model.

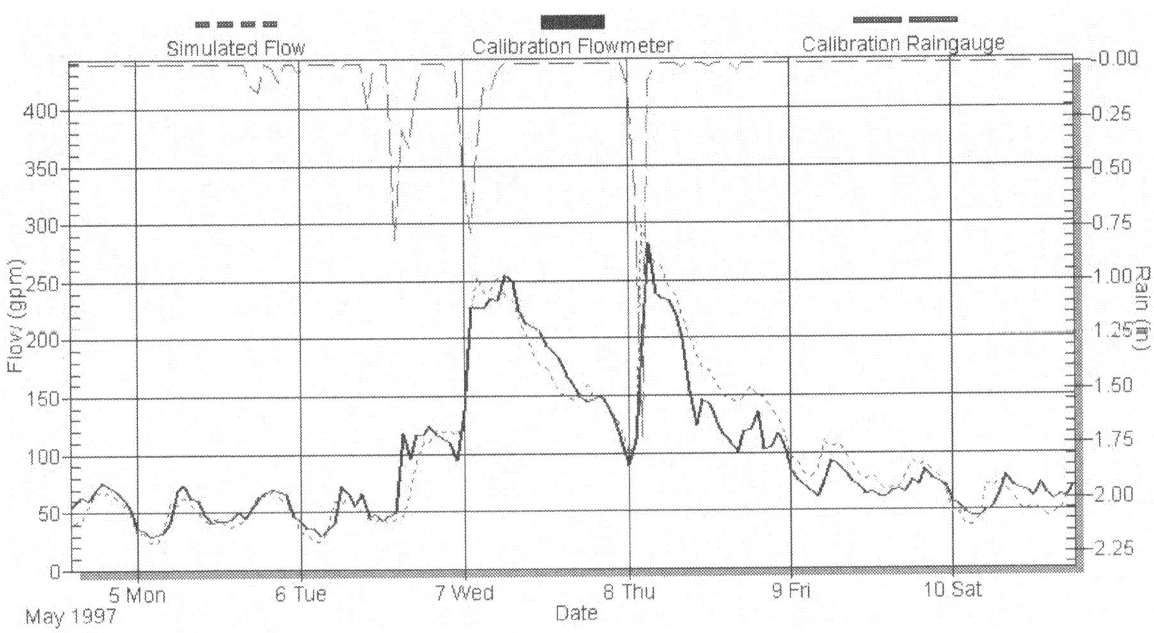

Figure B-6. Meter 2-203 (Basin A) Post-Rehabilitation Simulated with Pre-Rehabilitation Model Coefficients.

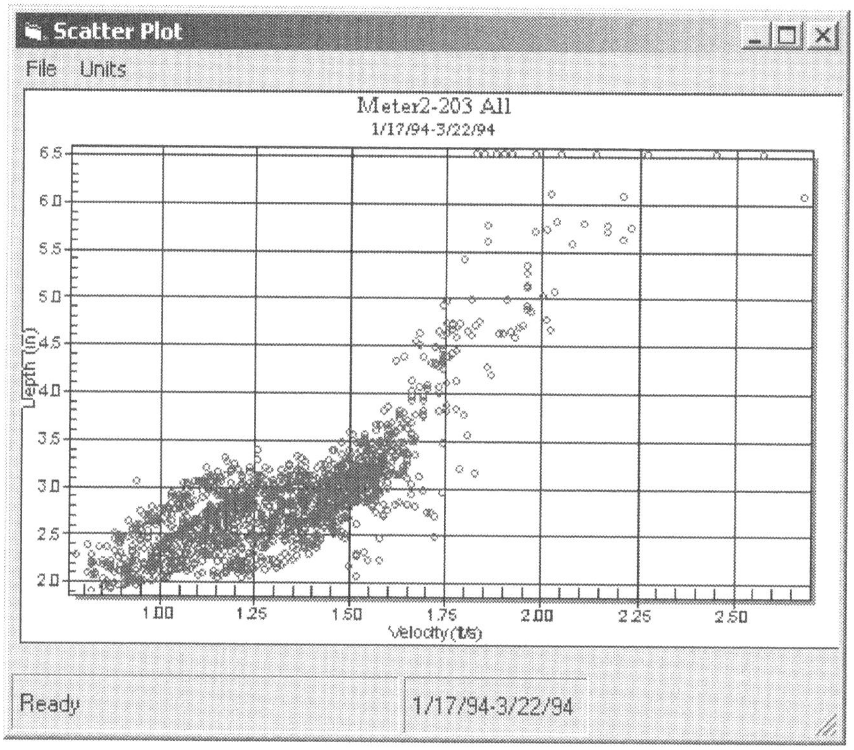

Figure B-7. Meter 2-203 (Basin A) Pre-Rehabilitation Velocity-Depth Scatter Plot.

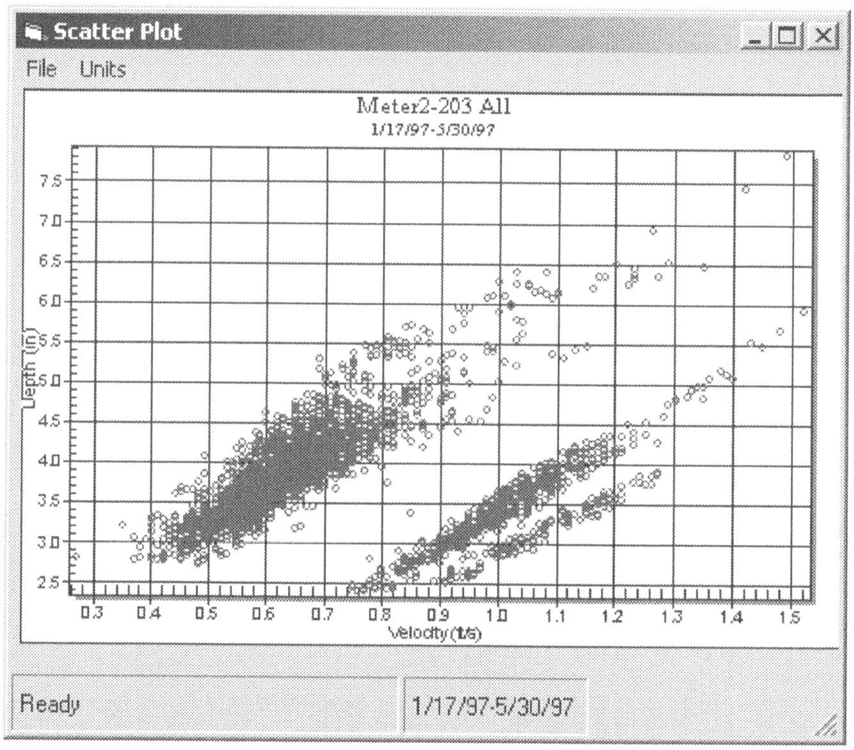

Figure B-8. Meter 2-203 (Basin A) Post-Rehabilitation Velocity-Depth Scatter Plot.

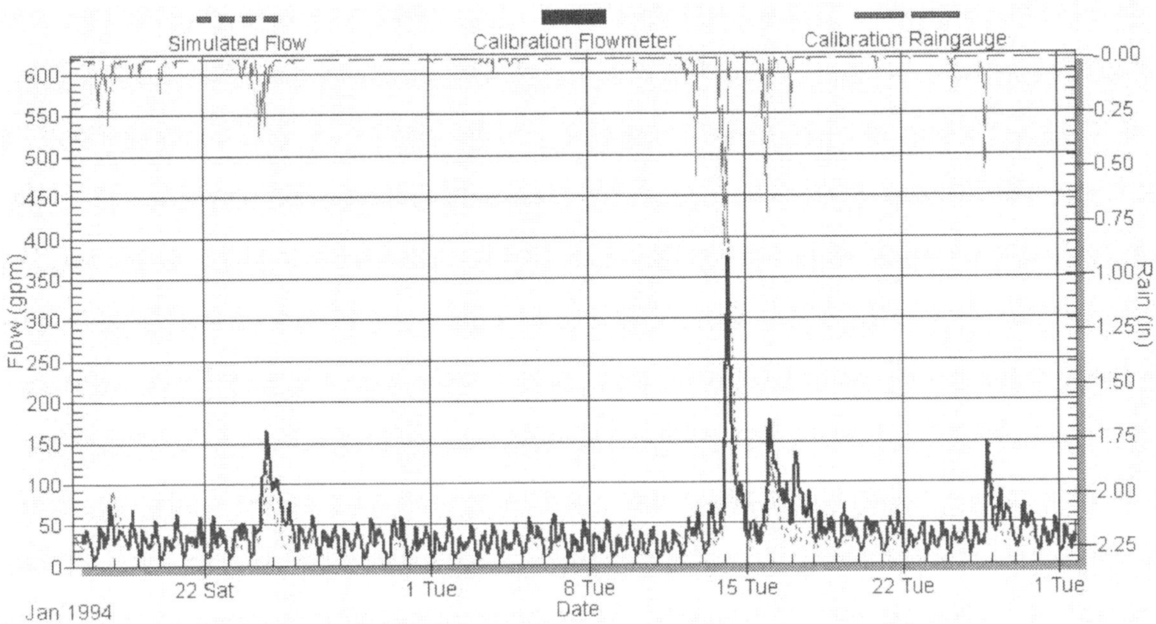

Figure B-9. Meter 2-205 (Basin B) Pre-Rehabilitation Period: January 1994.

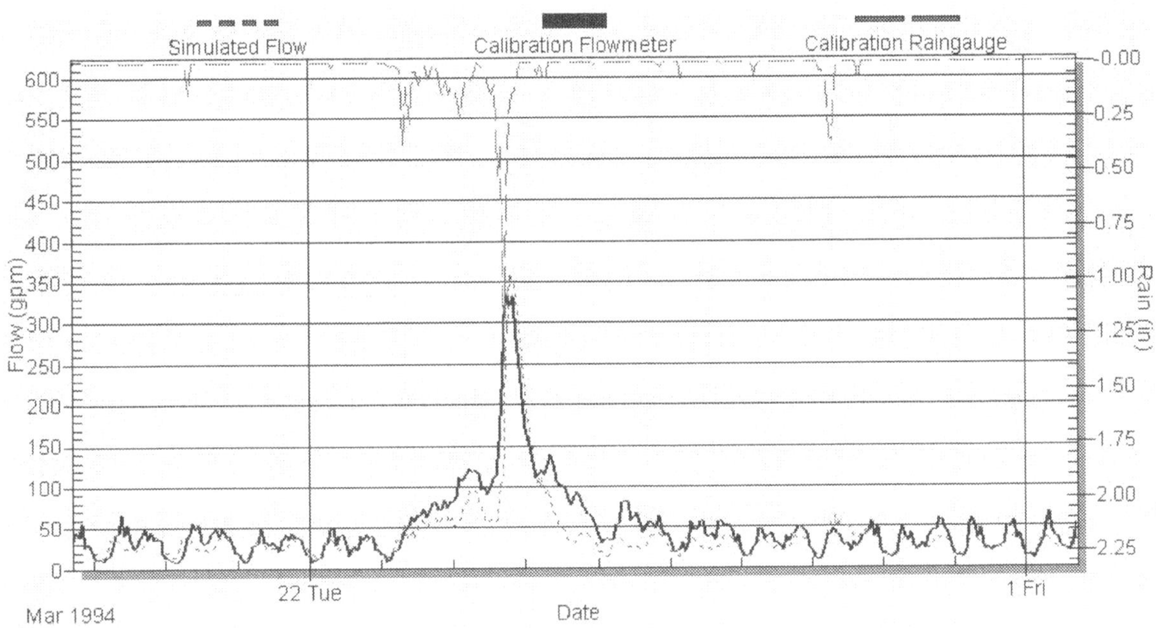

Figure B-10. Meter 2-205 (Basin B) Pre-Rehabilitation Period: March 1994.

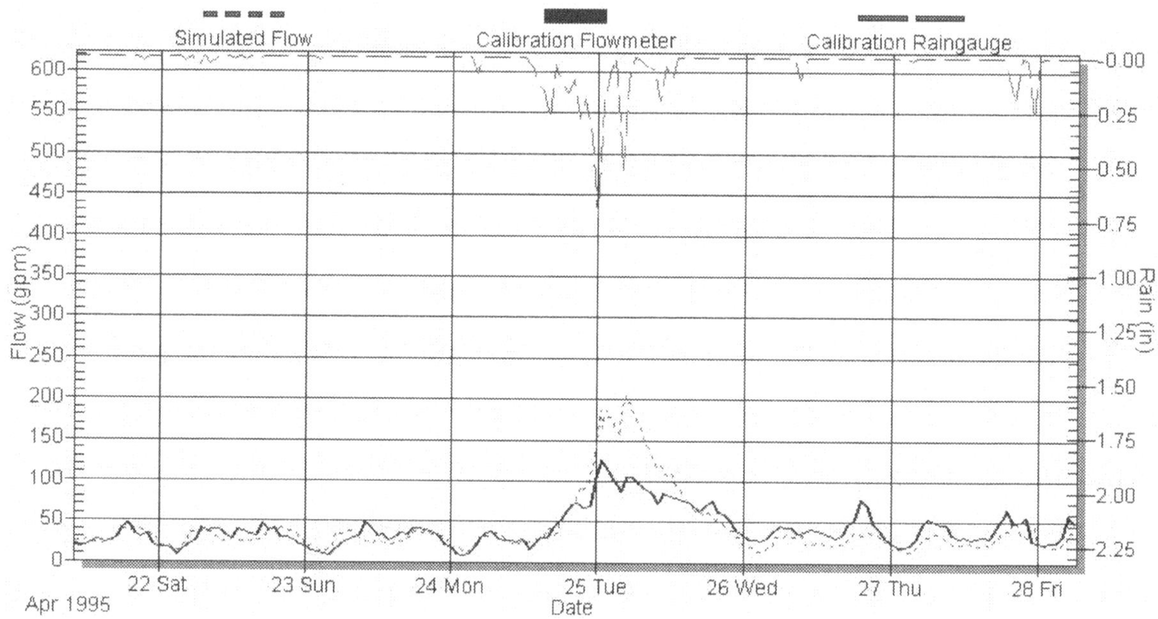

Figure B-11. Meter 2-205 (Basin B) Pre-Rehabilitation Period: April 1995.

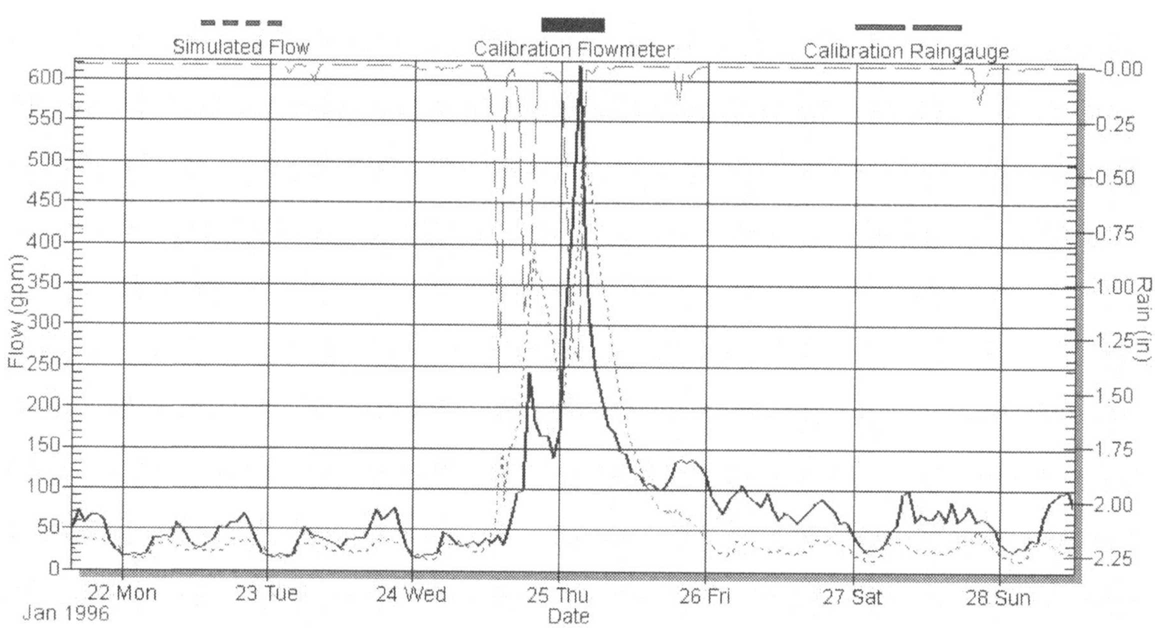

Figure B-12. Meter 2-205 (Basin B) Pre-Rehabilitation Period: January 1996.

Water Environment Research Foundation

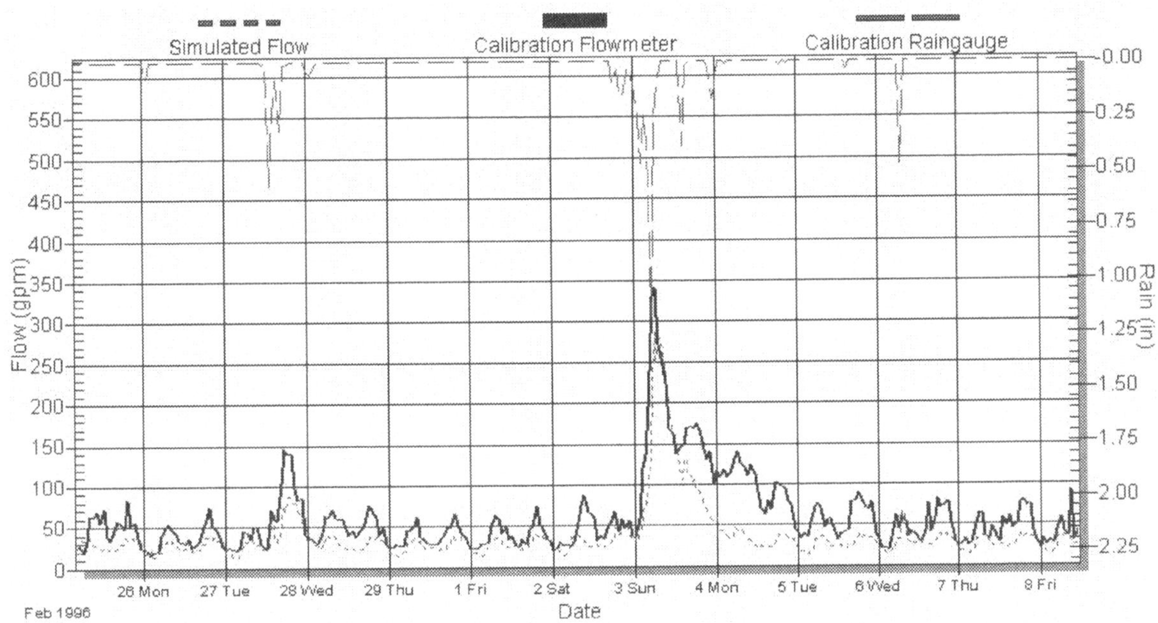

Figure B-13. Meter 2-205 (Basin B) at Start of Rehabilitation Project.
(Construction began February 29, 1996. Model results shown use pre-rehabilitation calibration parameters.)

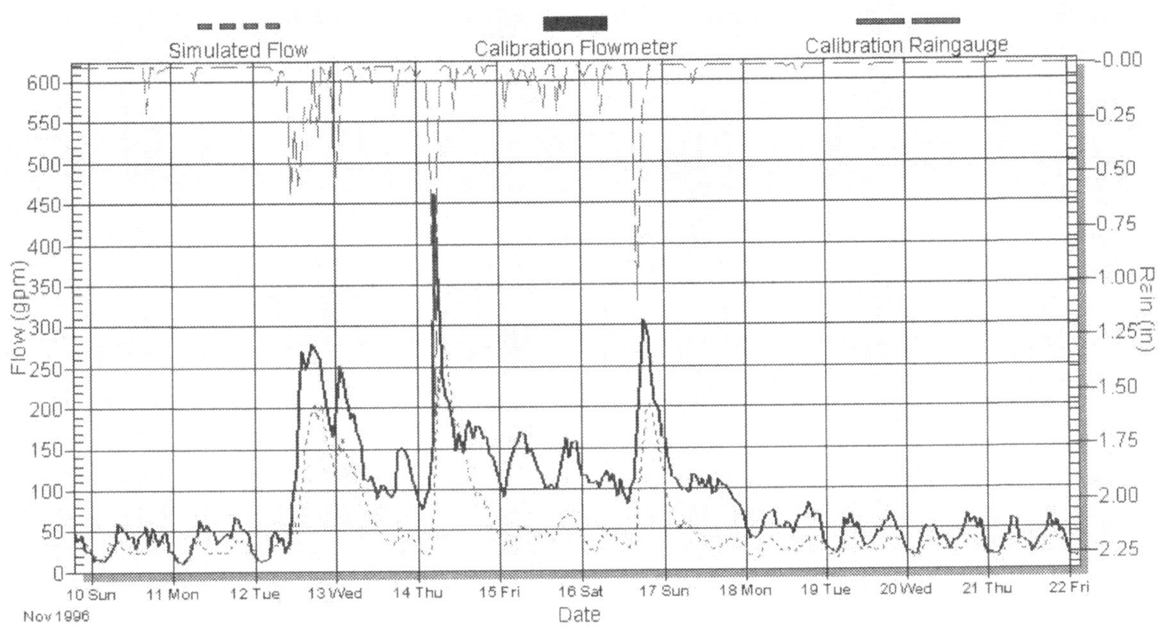

Figure B-14. Meter 2-205 (Basin B) Period Immediately Following Completion of Construction: October 31, 1996.
(Model results shown use pre-rehabilitation calibration parameters.)

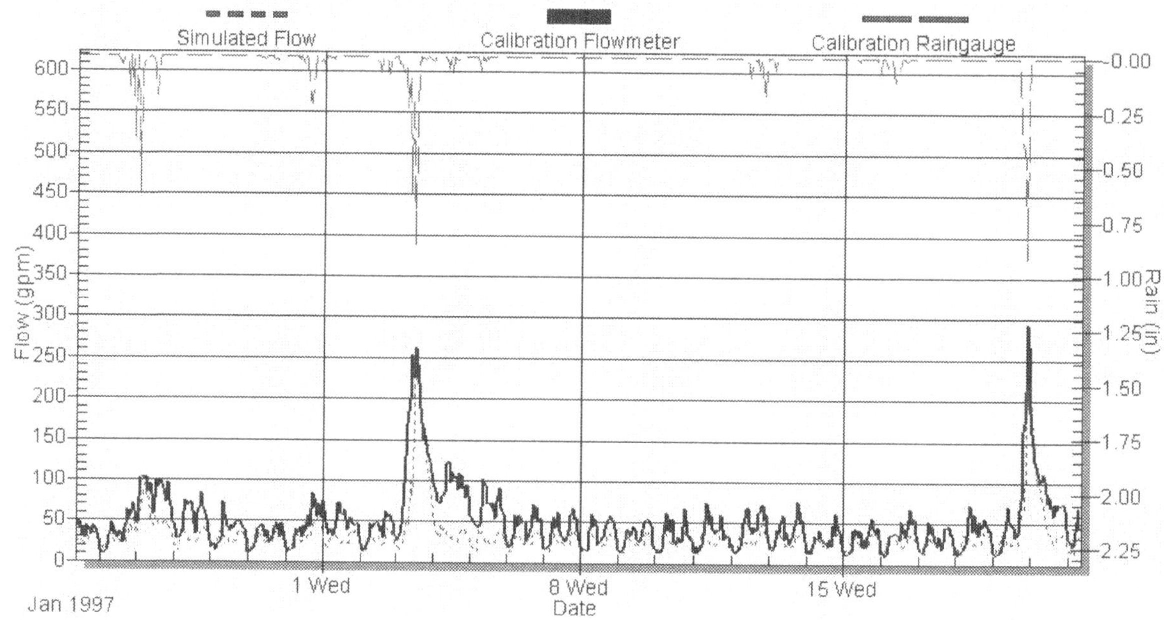

Figure B-15. Meter 2-205 (Basin B) Post-Rehabilitation Period: January 1997.
(Model results shown use pre-rehabilitation calibration parameters.)

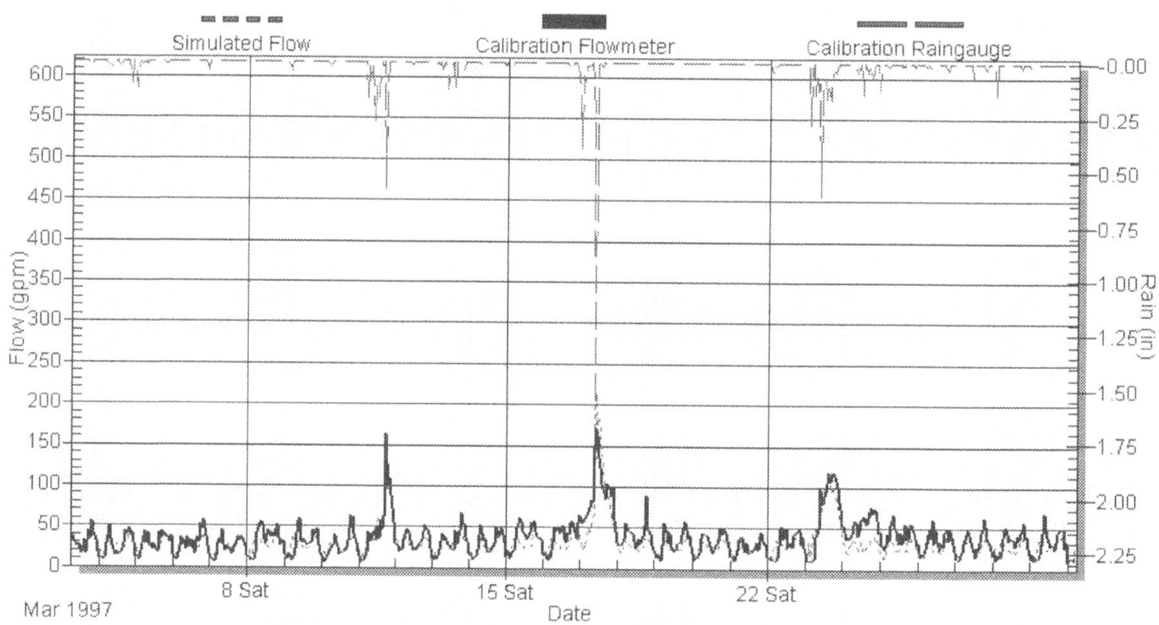

Figure B-16. Meter 2-205 (Basin B) Post-Rehabilitation Period: March 1997.
(Model results shown use pre-rehabilitation calibration parameters.)

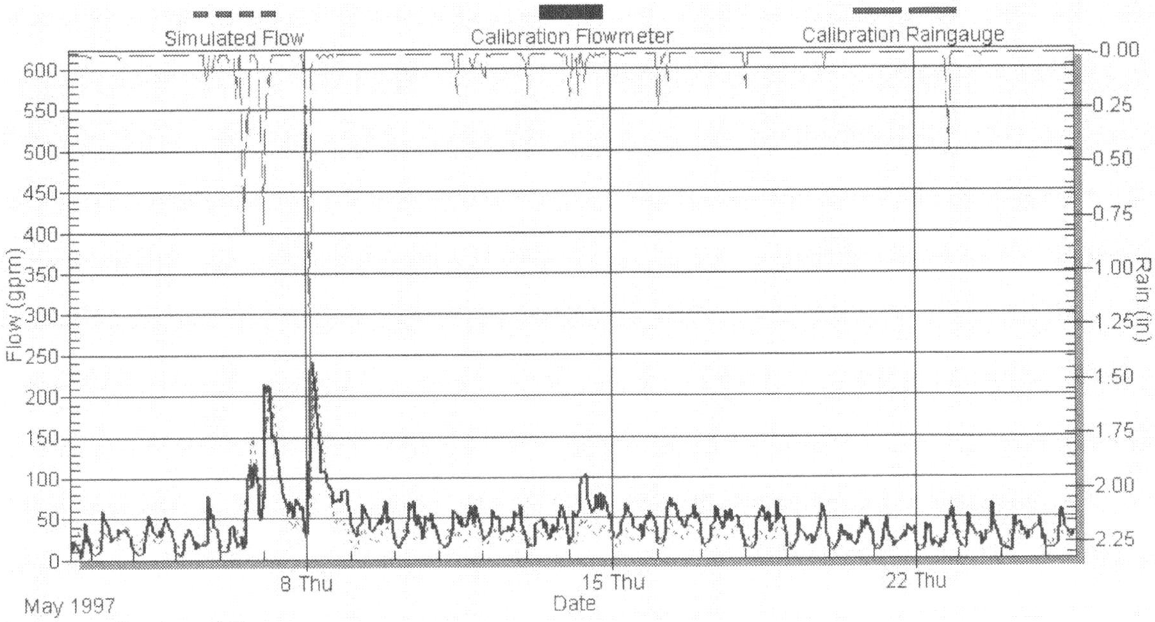

Figure B-17. Meter 2-205 (Basin B) Post-Rehabilitation Period: May 1997.
(Model results shown use pre-rehabilitation calibration parameters.)

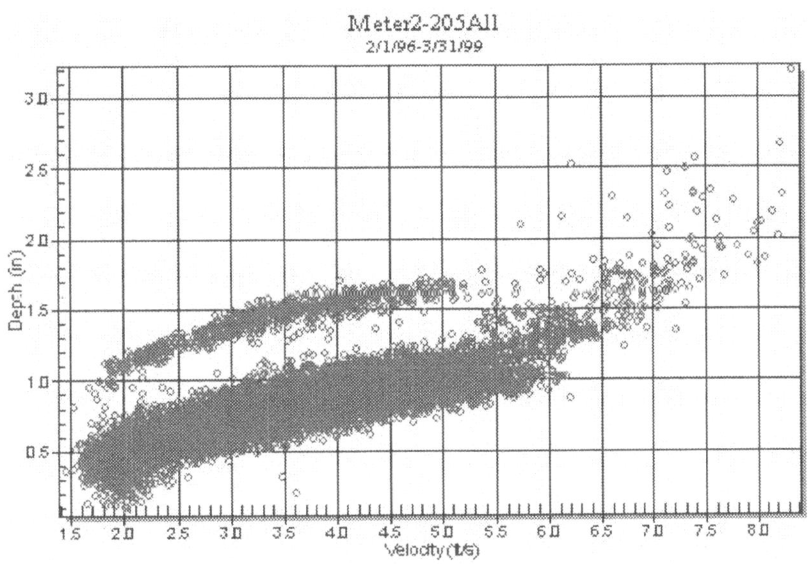

Figure B-18. Meter 2-205 (Basin B) Velocity-Depth Scatter Plot.
(Cluster of data above main group is from single month with suspicious data. This was not used in calibration.)

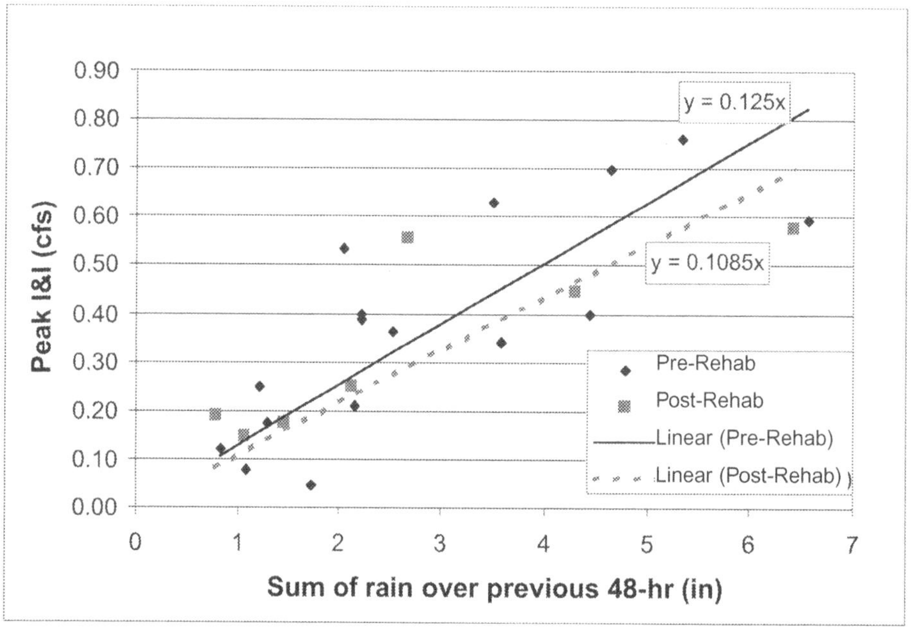

Figure B-19. Basin A Q vs. RDII Plot (Peak Measured Peak RDII vs. Sum of Rainfall Occurring in Previous 48 Hours).

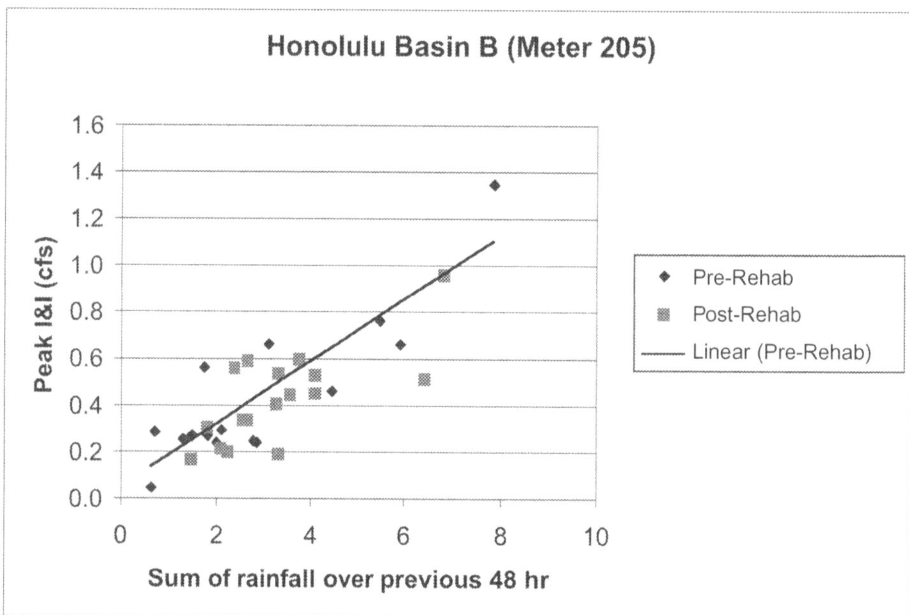

Figure B-20. Basin B Q vs. RDII Plot (Peak Measured Peak RDII vs. Sum of Rainfall Occurring in Previous 48 Hours).

LOTT CASE STUDY:

CALIBRATIONS

Table C-1. OL22 and OL24 Calibration Periods and Approximate Error.

Basin	Calibration Period	Calibration Error		Description
		Total Volume[1]	Peak Hour	
OL22	Aug. 1, 1996–Apr. 30, 1997	0.6%	9.3%[2]	Before 1997 demonstration project #1
	Oct. 1, 1997–Apr. 30, 1998	2.0%	2.2%[3]	After 1997 demonstration project #1
	Sept. 1, 1999–Apr. 23, 2000	1.5%	2.0%[4]	After 1999 demonstration project #2
OL24	Aug. 1, 1996–Apr. 30, 1997	1.0%	12.8%[5]	Before 1998 demonstration project
	Nov. 1, 1998–Apr. 1, 1999	3.1%	6%[6]	After 1998 demonstration project

[1]Average over calibration period.
[2]Occurred March 19, 1998.
[3]Occurred October 30, 1997.
[4]Occured December 15, 1999.
[5]Occurred March 19, 1998.
[6]Average of six events.

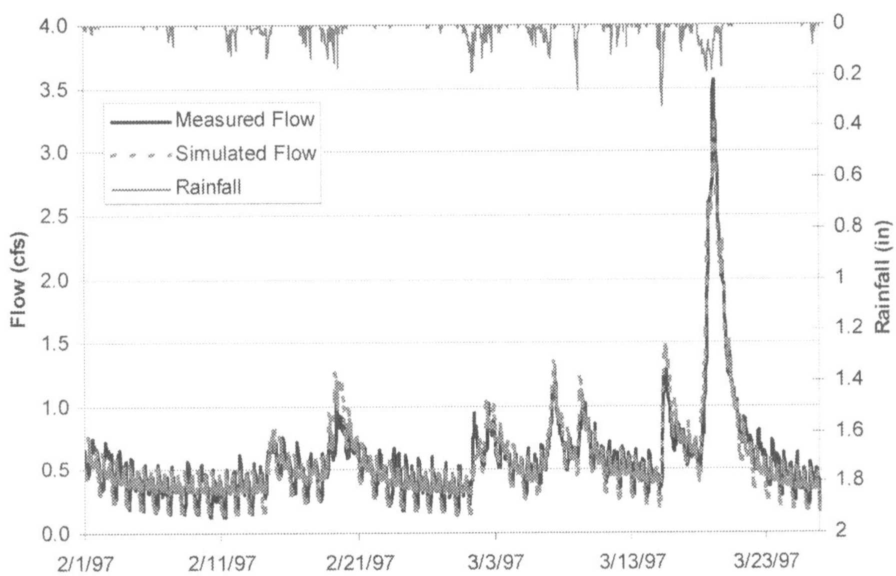

Figure C-1. OL22 Pre-rehabilitation Model Calibration Results: from February 1997 through April 1997.

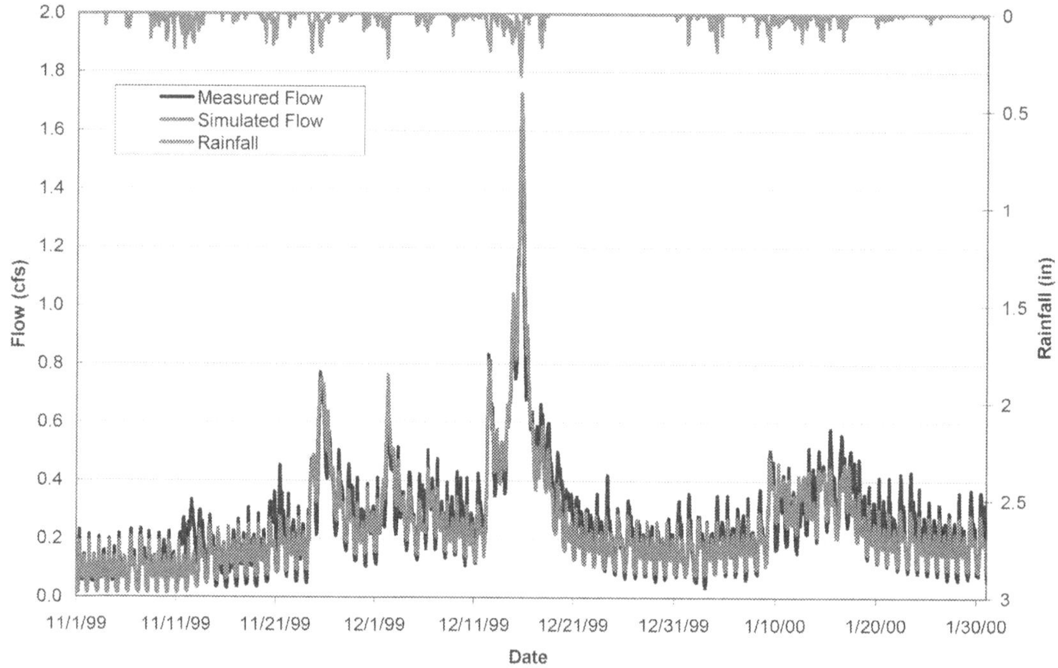

Figure C-2. OL22 Phase II Model Calibration Results: from November 1999 through January 2000.

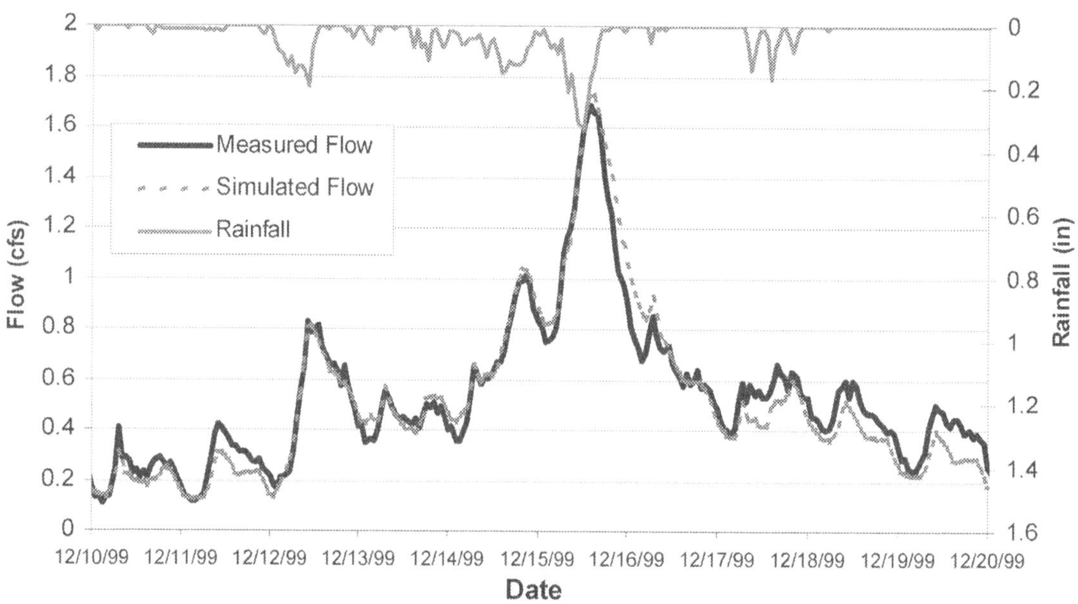

Figure C-3. OL22 Phase II Calibration for December 16, 1999-—Largest Flow Event of Wet Season.

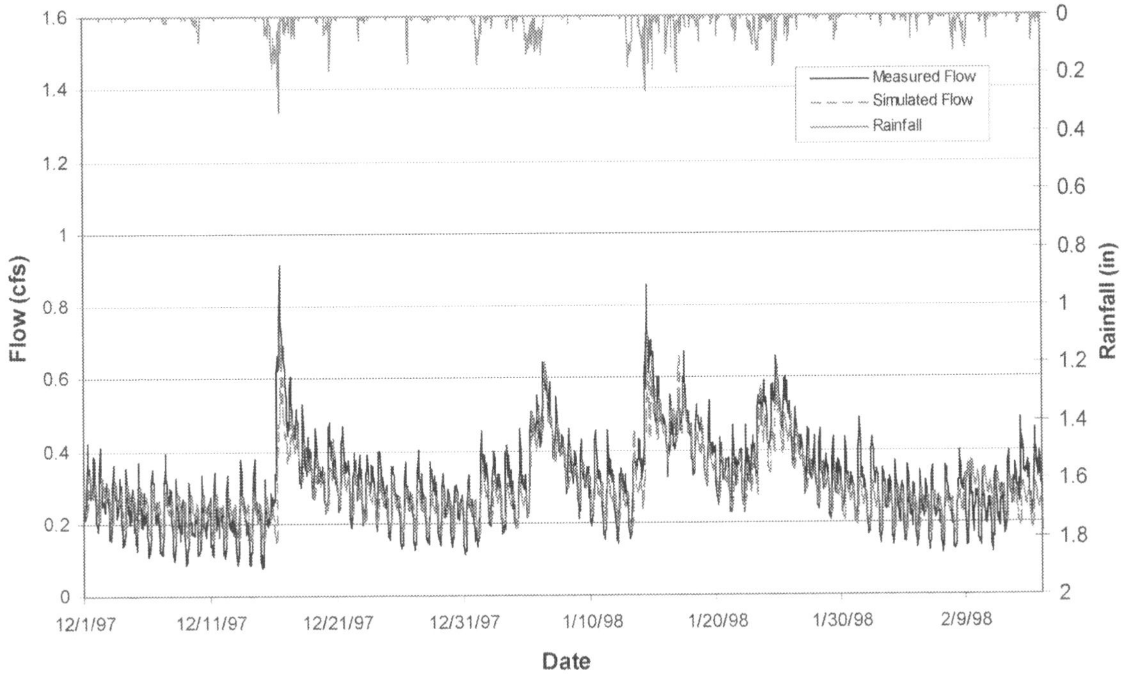

Figure C-4. OL24 Pre-Rehabilitation I/I Model Calibration: December 1997 through February 1998.

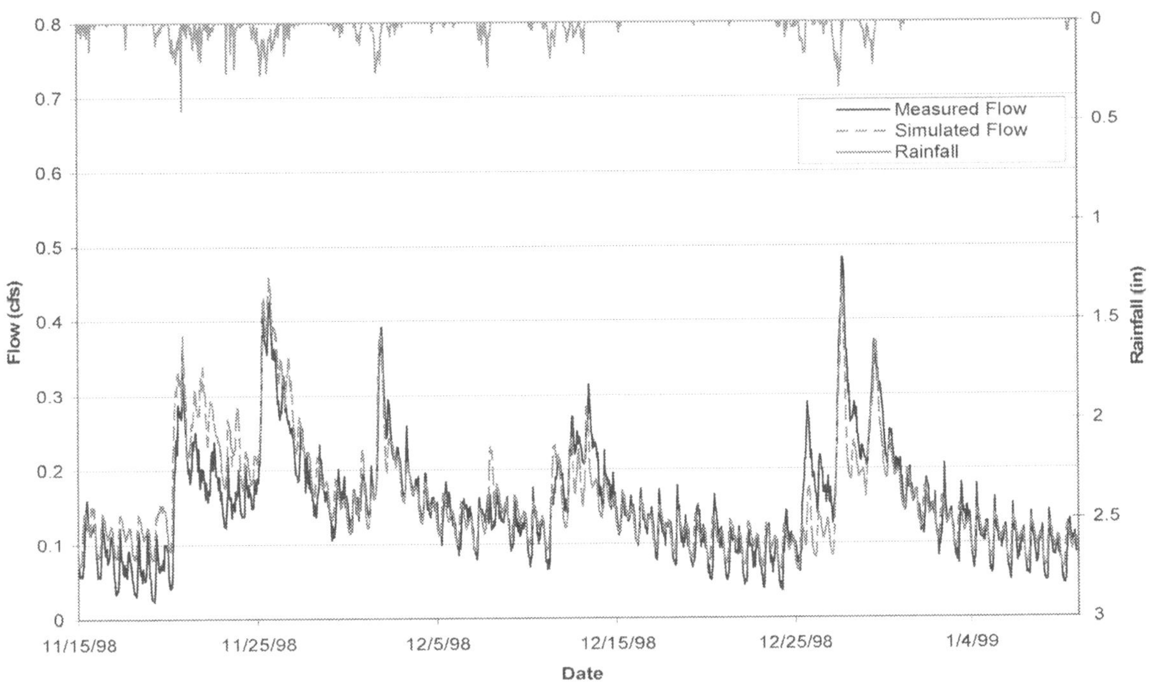

Figure C-5. OL24 Post-Rehabilitation I/I Model Calibration: November 1998 through January 1999.

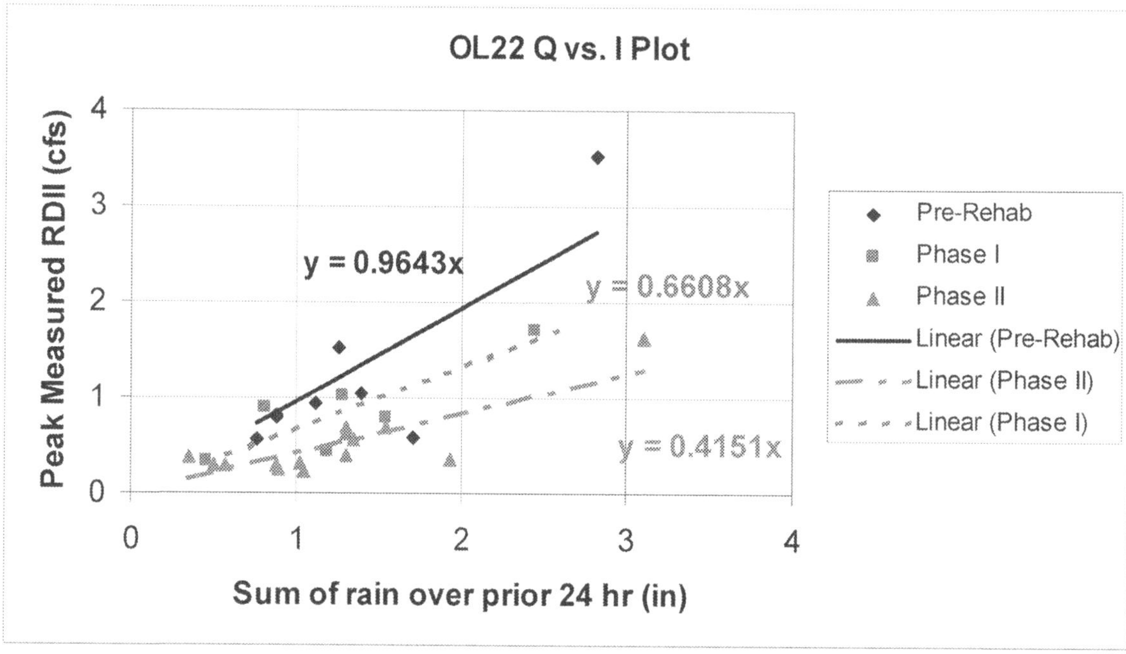

Figure C-6. OL22 Peak Measured RDII versus vs. Sum of Rainfall in the Previous 24 Hours.

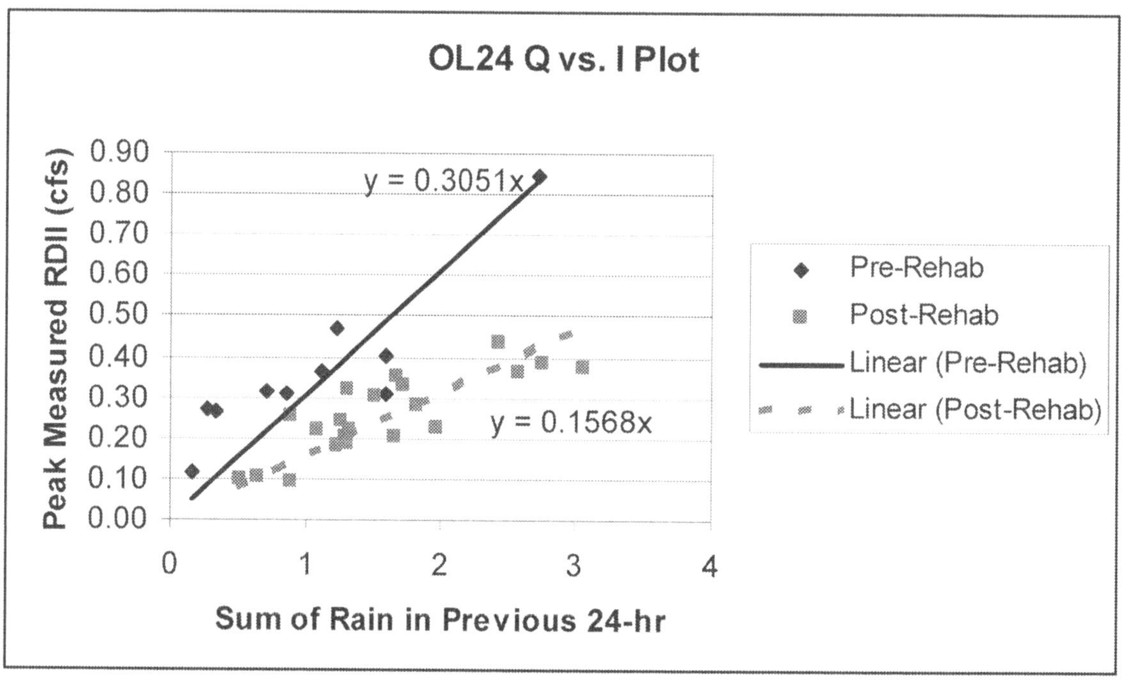

Figure C-7. OL24 Peak Measured RDII versus vs. Sum of Rainfall in the Previous 24 Hours.

King County Case Study: Supplementary Information

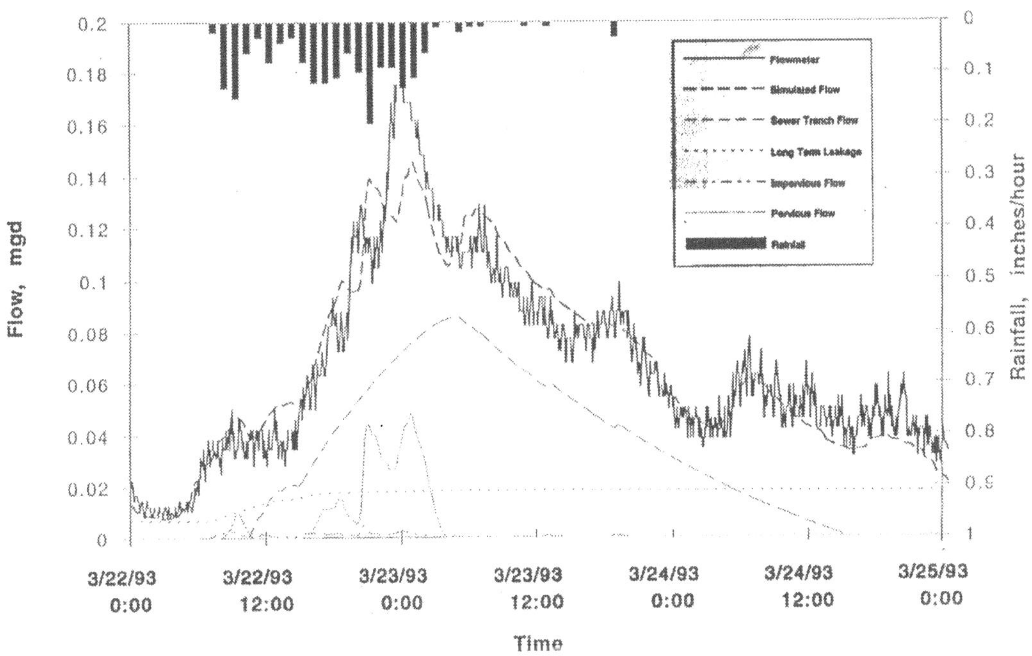

Figure D-1. March 22–25, 1993: King County Kent Control Basin Model Calibration, 25 Acres.

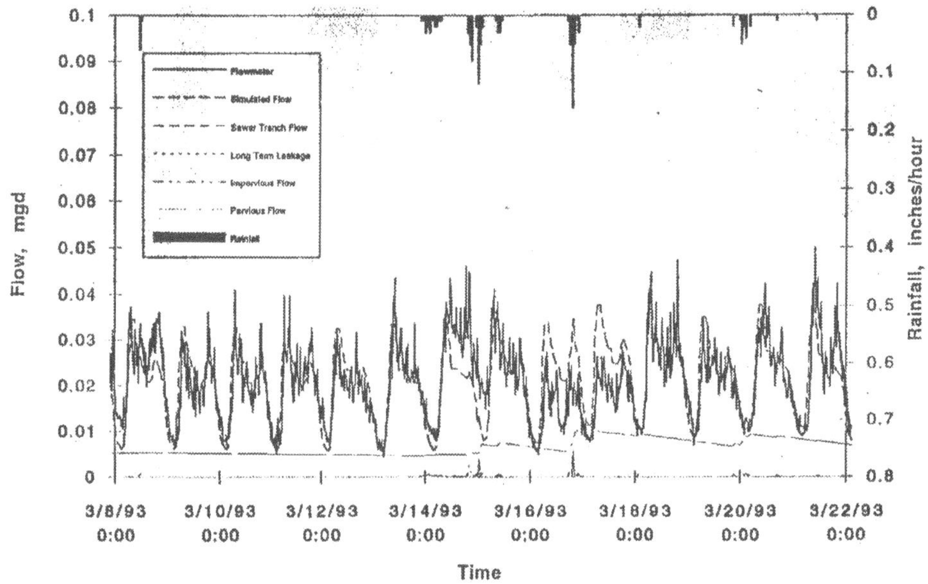

Figure D-2. March 8–22, 1993: King County Kent Control Basin Model Calibration, 25 Acres.

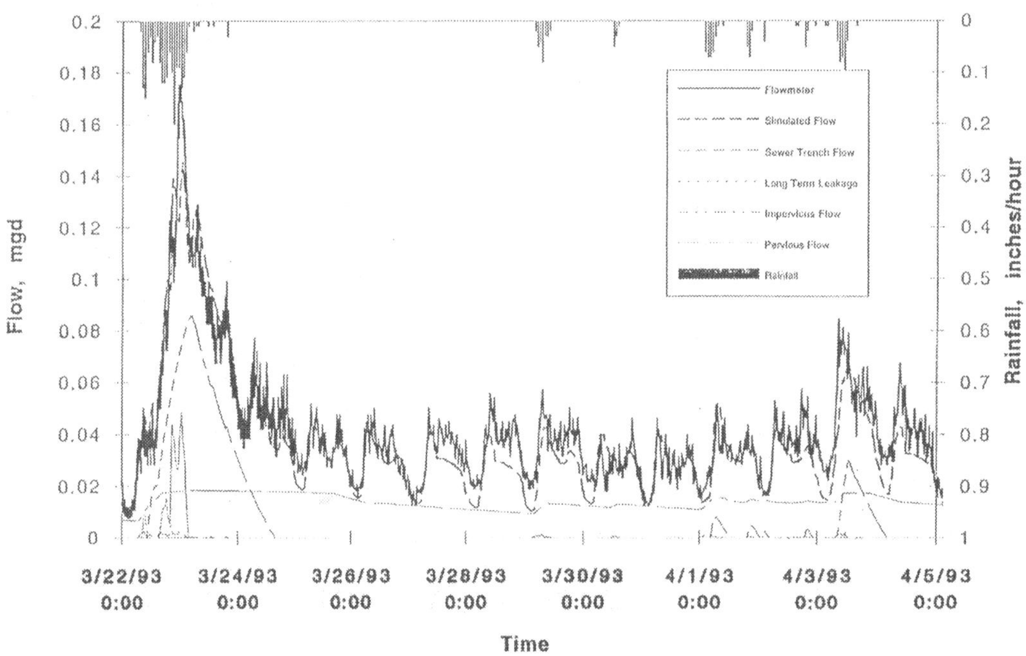

Figure D-3. March 22–April 5, 1993: King County Kent Control Basin Model Calibration, 25 Acres.

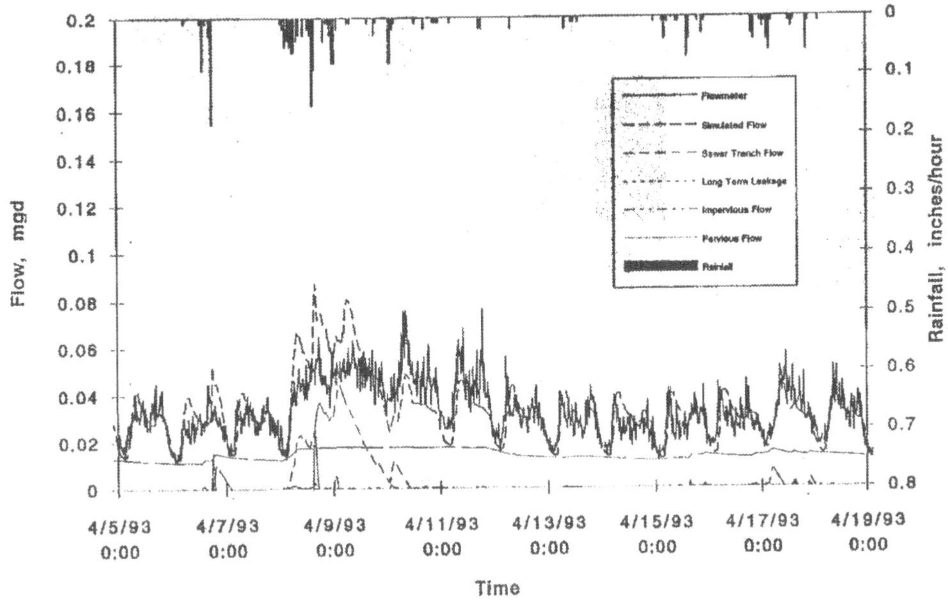

Figure D-4. April 5–19, 1993: King County Kent Control Basin Model Calibration, 25 Acres.

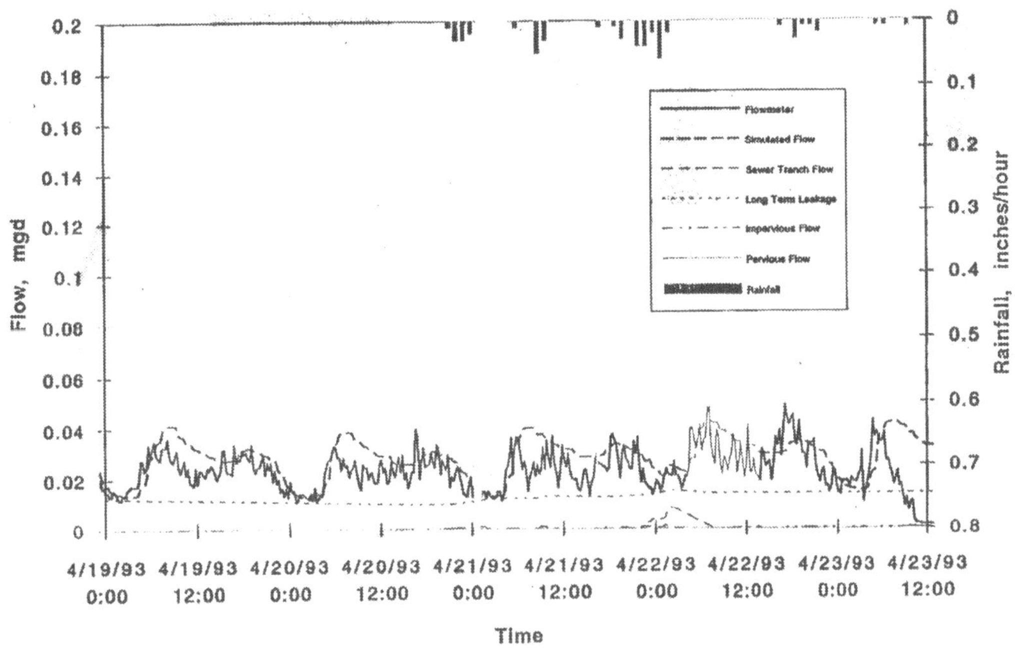

Figure D-5. March 19–23, 1993: King County Kent Control Basin Model Calibration, 25 Acres.

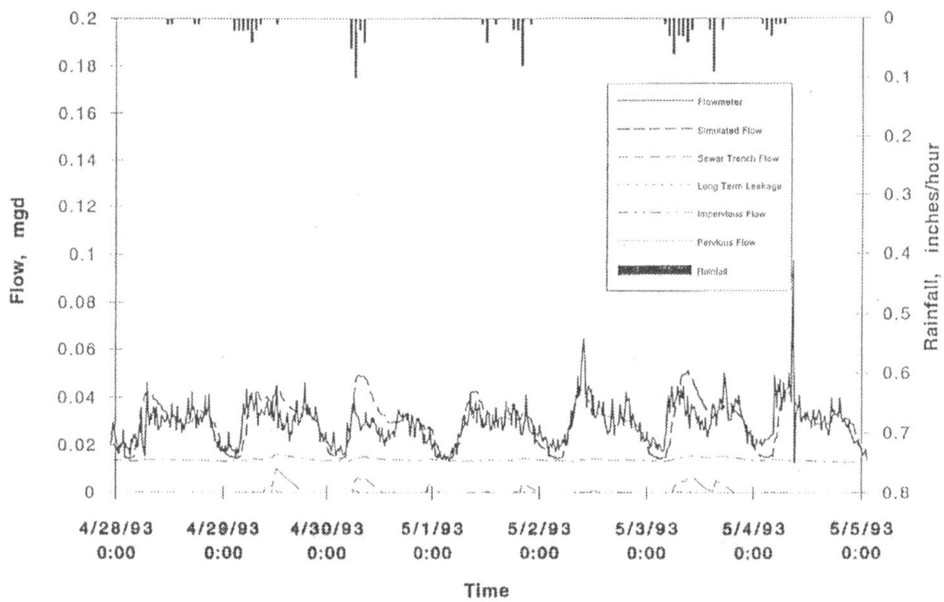

Figure D-6. April 28–May 5, 1993: King County Kent Control Basin Model Calibration, 25 Acres.

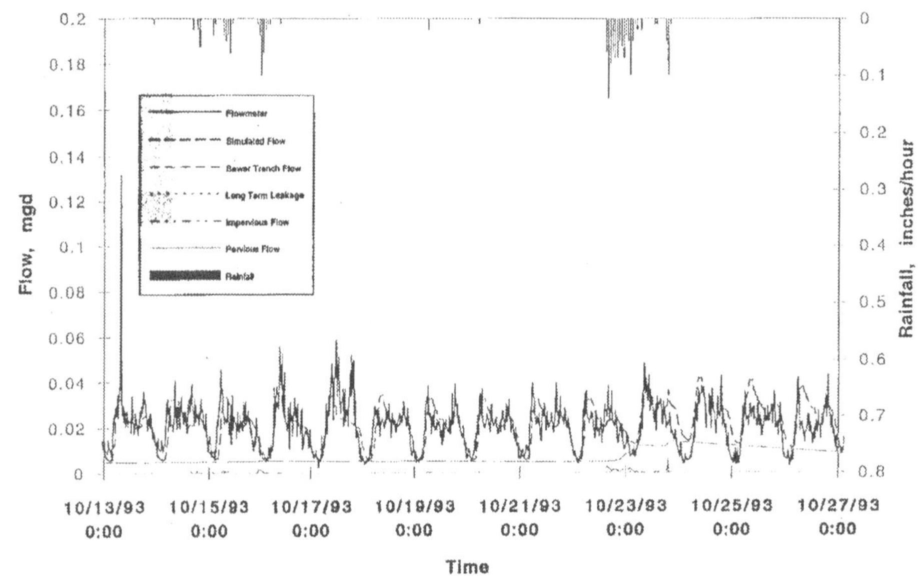

Figure D-7. October 13–27, 1993: King County Kent Control Basin Model Calibration, 25 Acres.

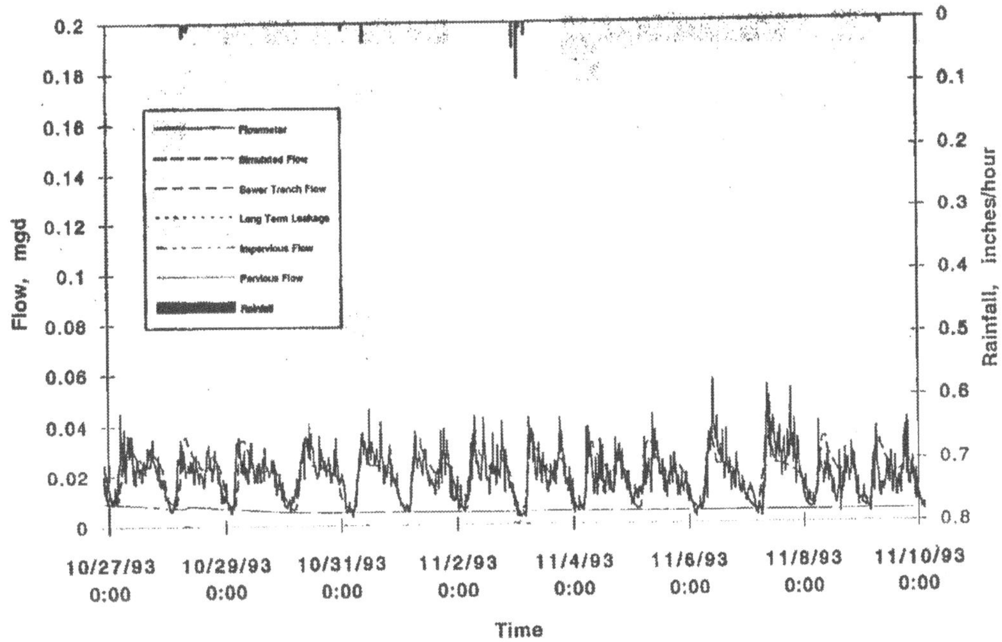

Figure D-8. October 27–November 10, 1993: King County Kent Control Basin Model Calibration, 25 Acres.

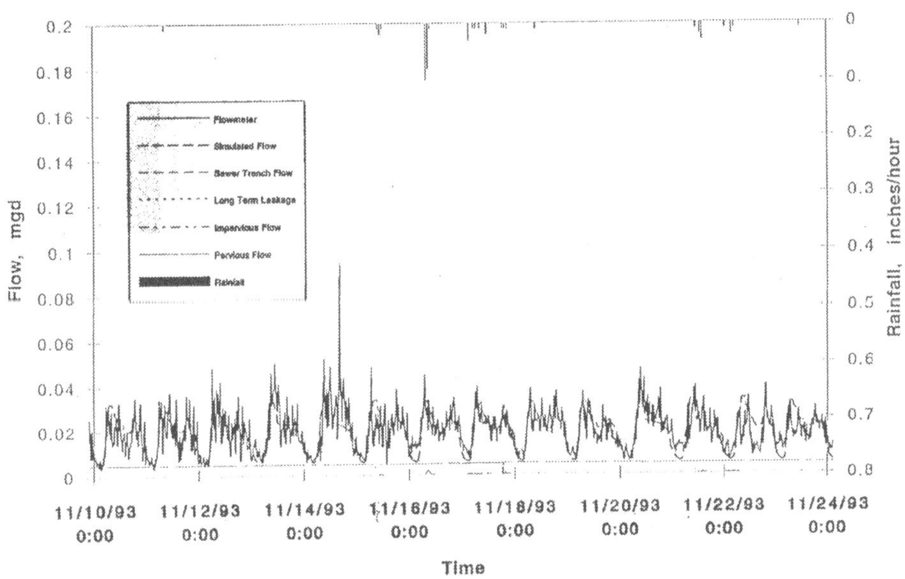

Figure D-9. November 10–24, 1993: King County Kent Control Basin Model Calibration, 25 Acres.

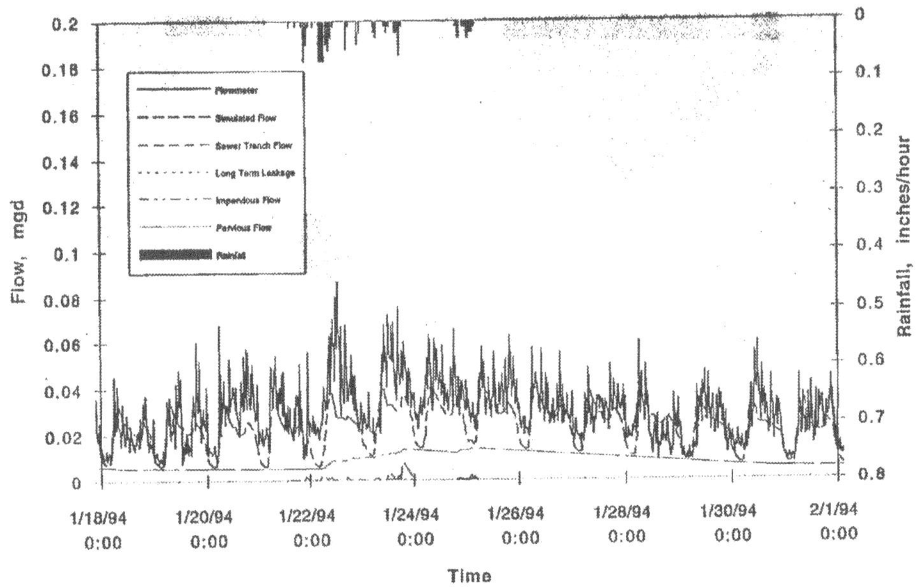

Figure D-10. January 18–February 1, 1994: King County Kent Control Basin Model Calibration, 25 Acres.

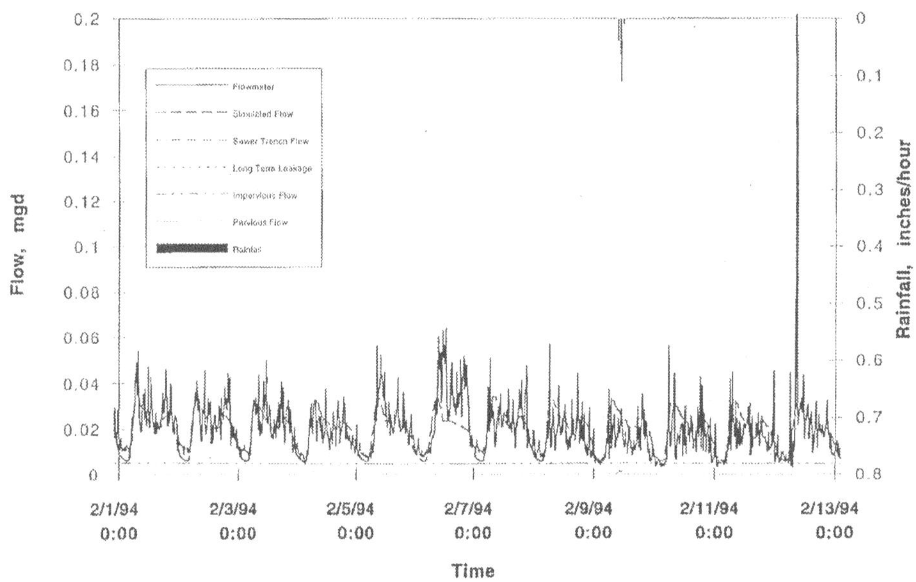

Figure D-11. February 1–13, 1994: King County Kent Control Basin Model Calibration, 25 Acres.

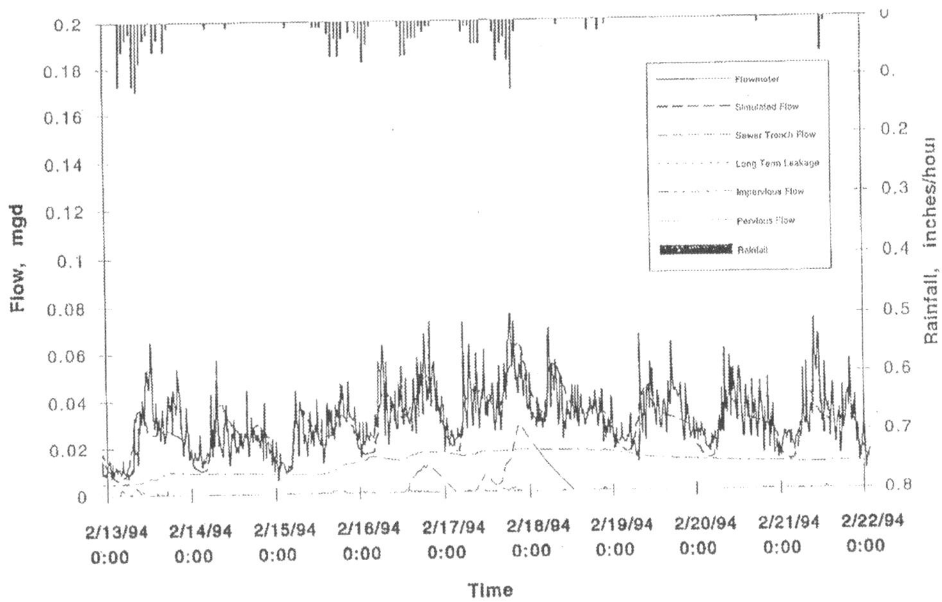

Figure D-12. February 13–22, 1994: King County Kent Control Basin Model Calibration, 25 Acres.

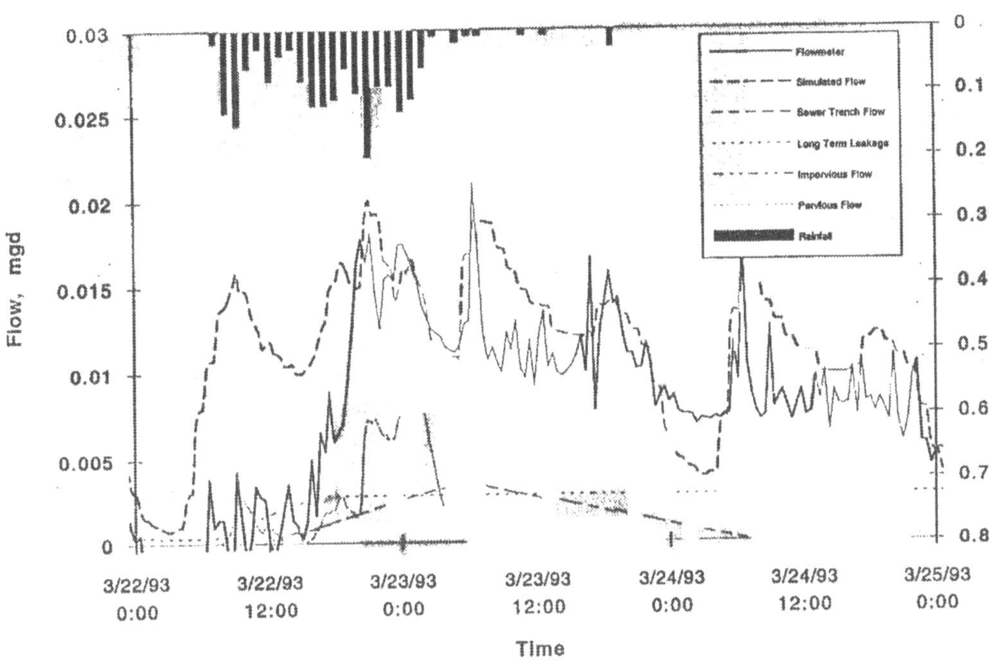

Figure D-13. March 22–25, 1993: King County Kent Rehabilitation Basin Model Calibration, 10 Acres.

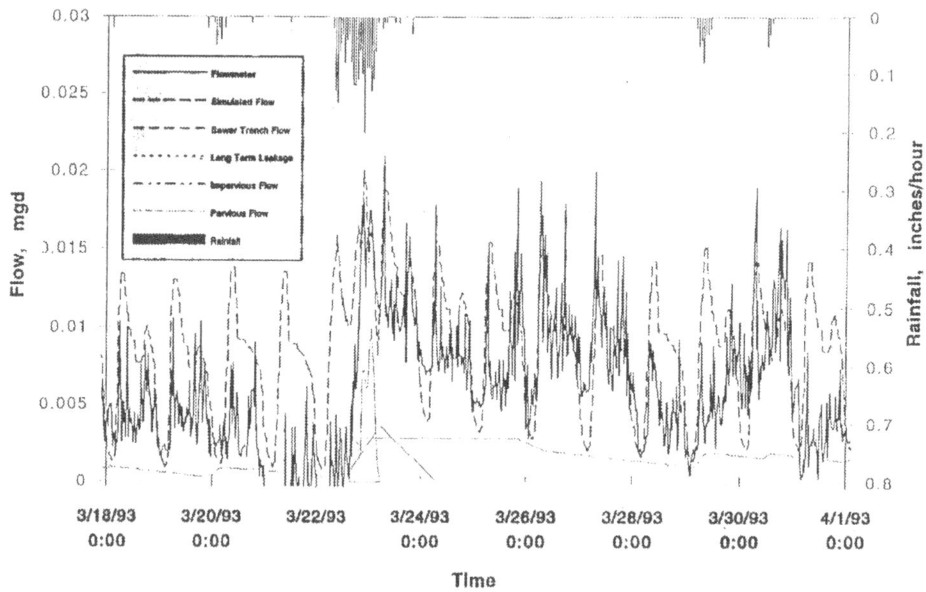

Figure D-14. March 18–April 1, 1993: King County Kent Rehabilitation Basin Model Calibration, 10 Acres.

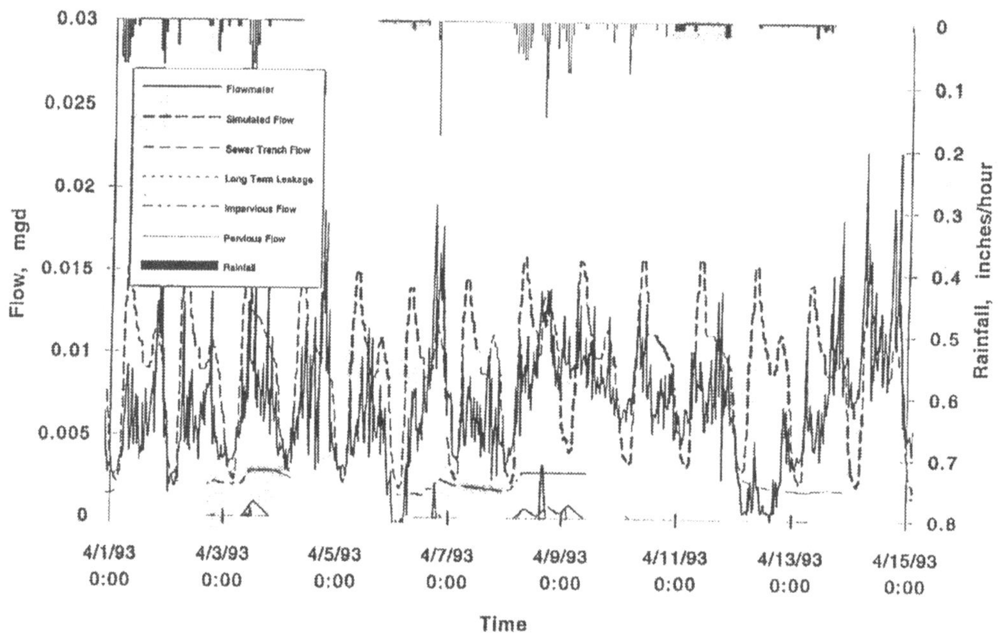

Figure D-15. April 1–15, 1993: King County Kent Rehabilitation Basin Model Calibration, 10 Acres.

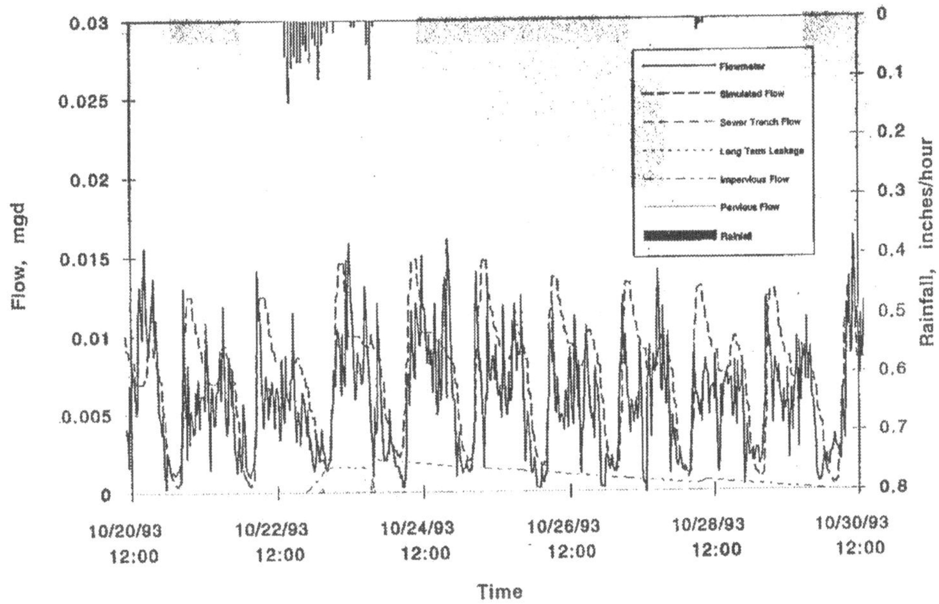

Figure D-16. October 20–30, 1993: King County Kent Rehabilitation Basin Model Calibration, 10 Acres.

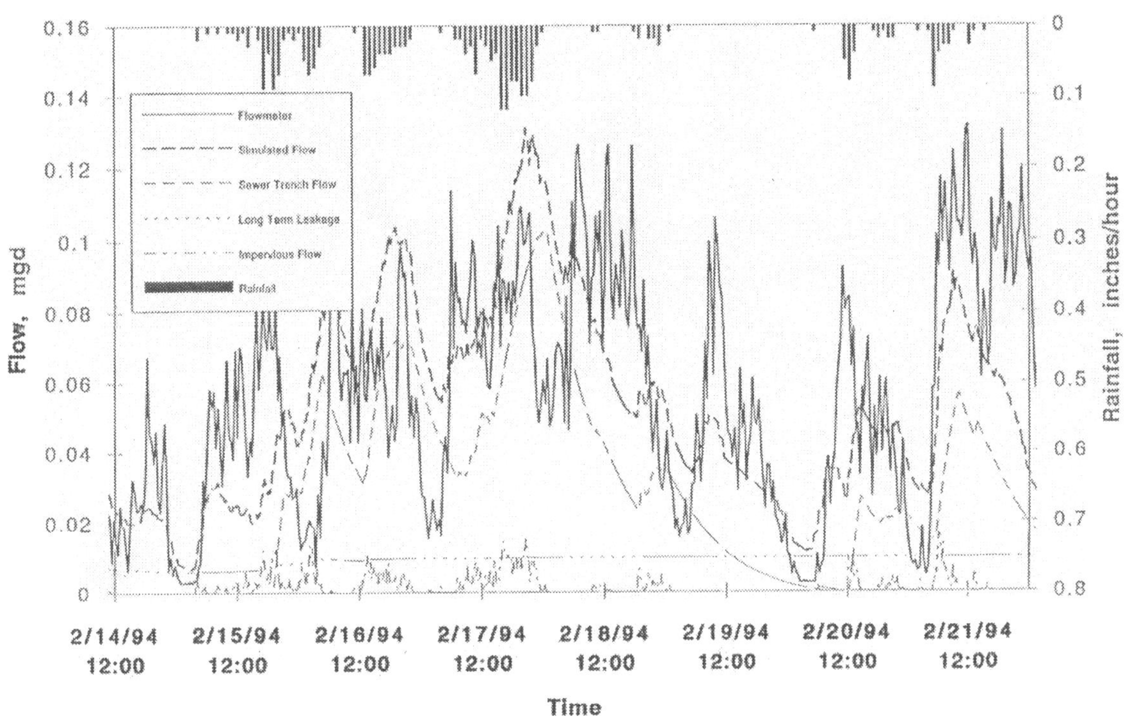

Figure D-17. February 1994: King County Issaquah Control Basin Model Calibration, 25 Acres.

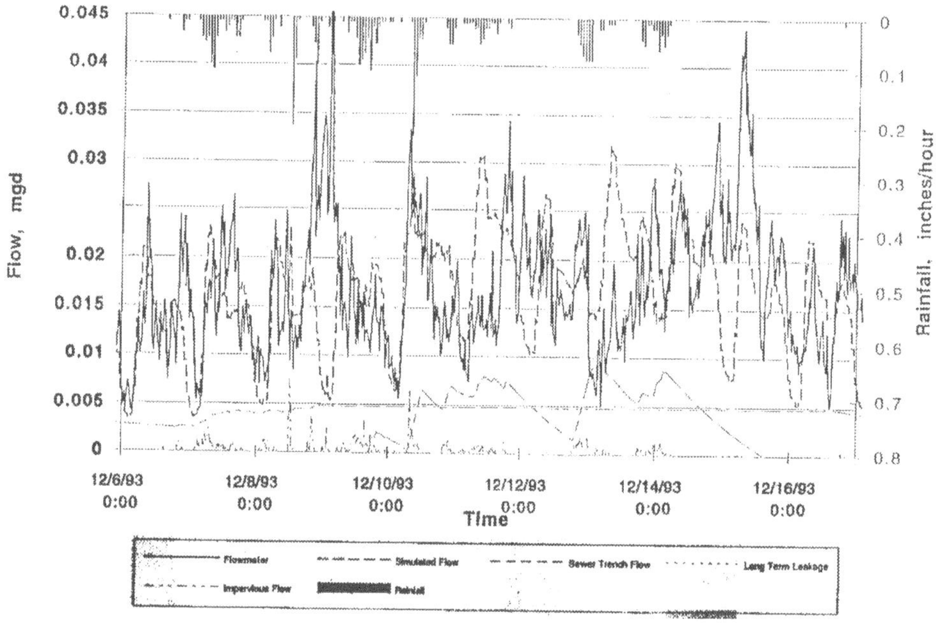

Figure D-18. December 1993: King County Issaquah Rehabilitation Basin Model Calibration, 10 Acres.

APPENDIX E

McCandless Case Study: Calibrations

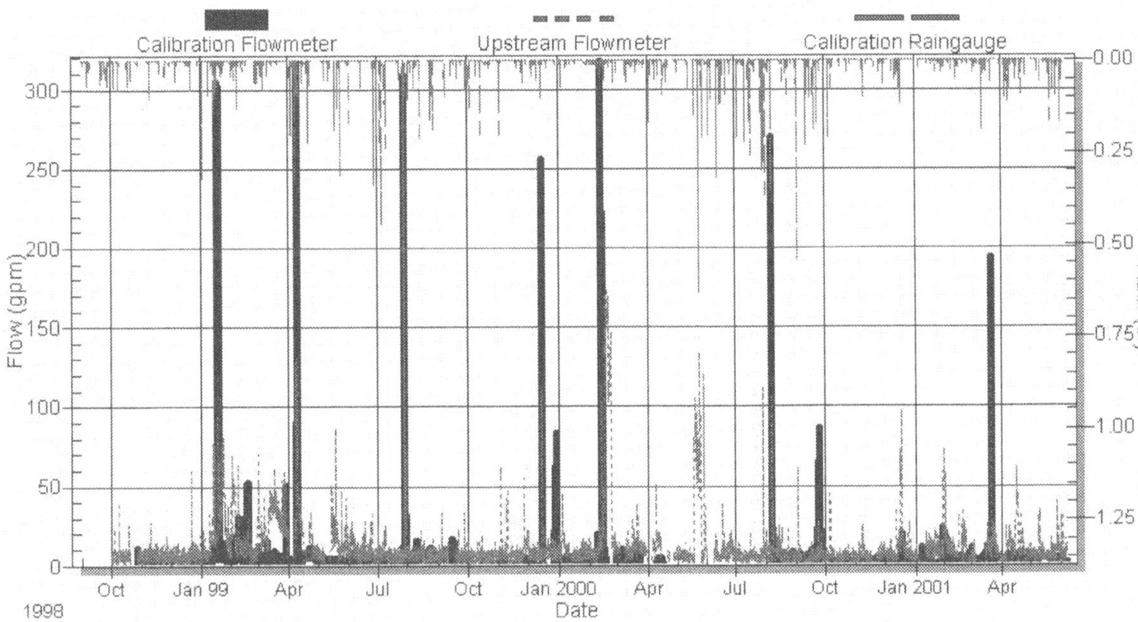

Figure E-1. Willoughby Road and Control Basin Comparison.
The Willoughby meter is shown as the thick line (labeled Calibration Flowmeter); control basin is shown as a thin dotted line (labeled Upstream Flowmeter); precipitation is plotted in a dashed line positive downward from the top of the chart.)

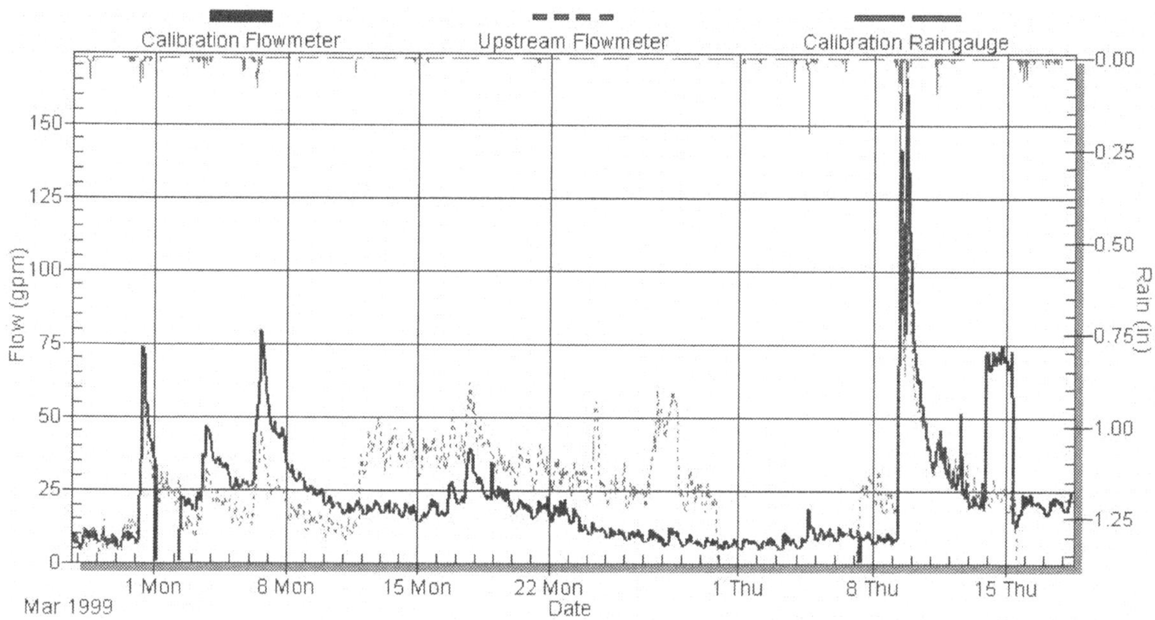

Figure E-2. Moreland Road (Inwood Meter) and Emerald Meter (Control Basin): March–April, 1999 Pre-Rehabilitation Data.

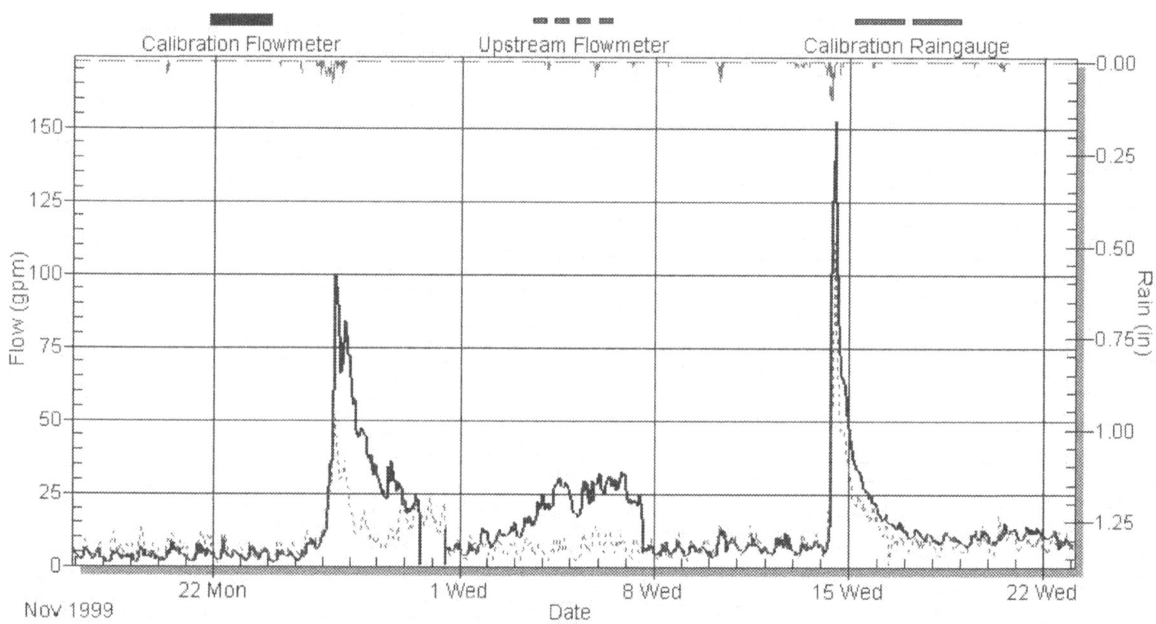

Figure E-3. Moreland Road (Inwood Meter) and Emerald Meter: November–December, 1999 Pre-Lateral Rehabilitation Data.

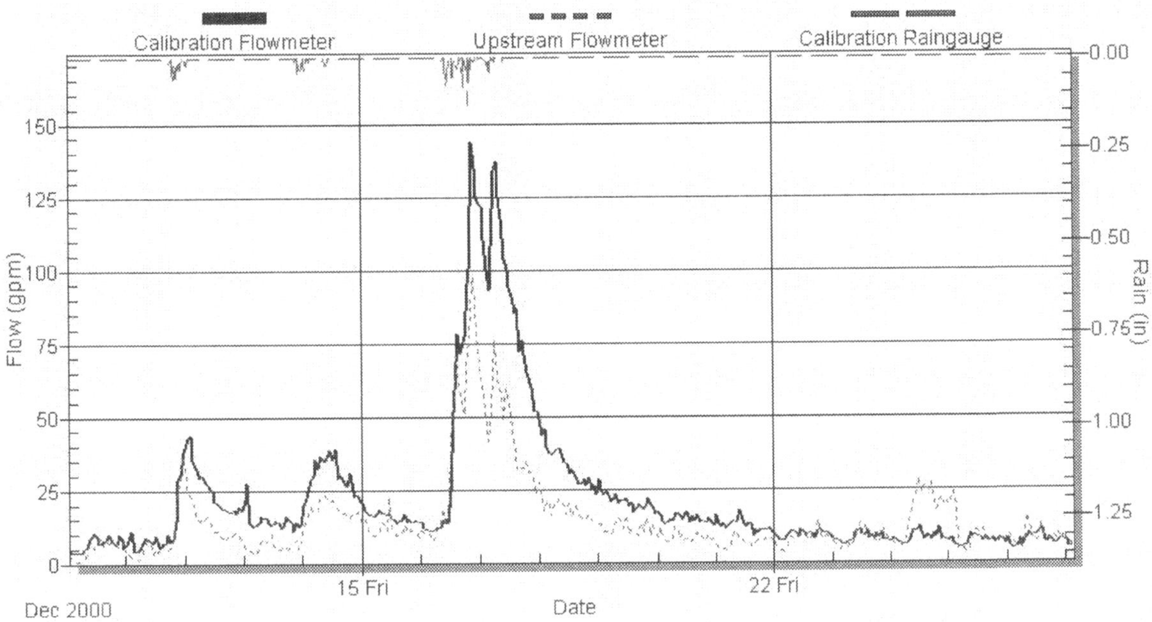

Figure E-4. Moreland Road (Inwood Meter) and Emerald Meter: December, 2000 Post-Rehabilitation Data.

MILWAUKEE CASE STUDY: CALIBRATIONS AND VELOCITY-DEPTH SCATTER PLOTS

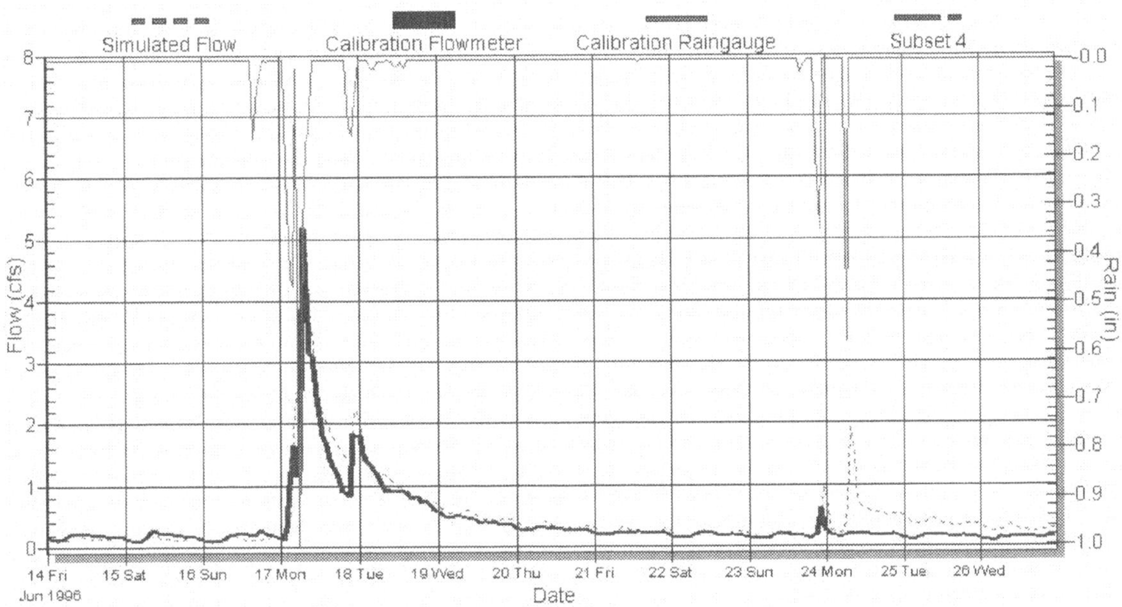

Figure F-1. Basin 1144: Pre-Rehabilitation Calibration.

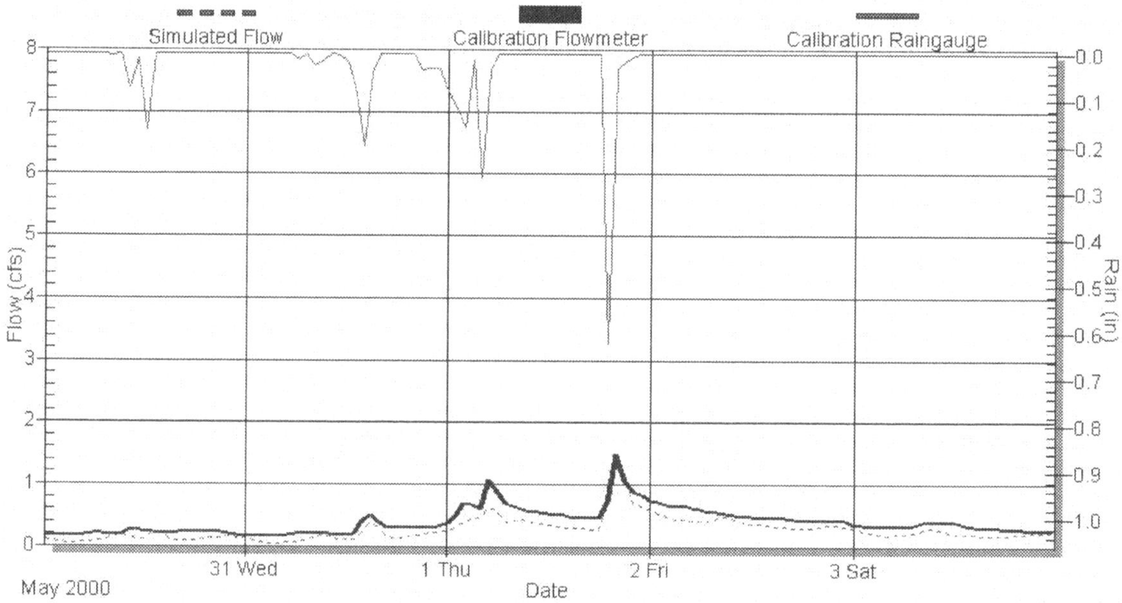

Figure F-2. Basin 1144: Post-Rehabilitation Calibration.

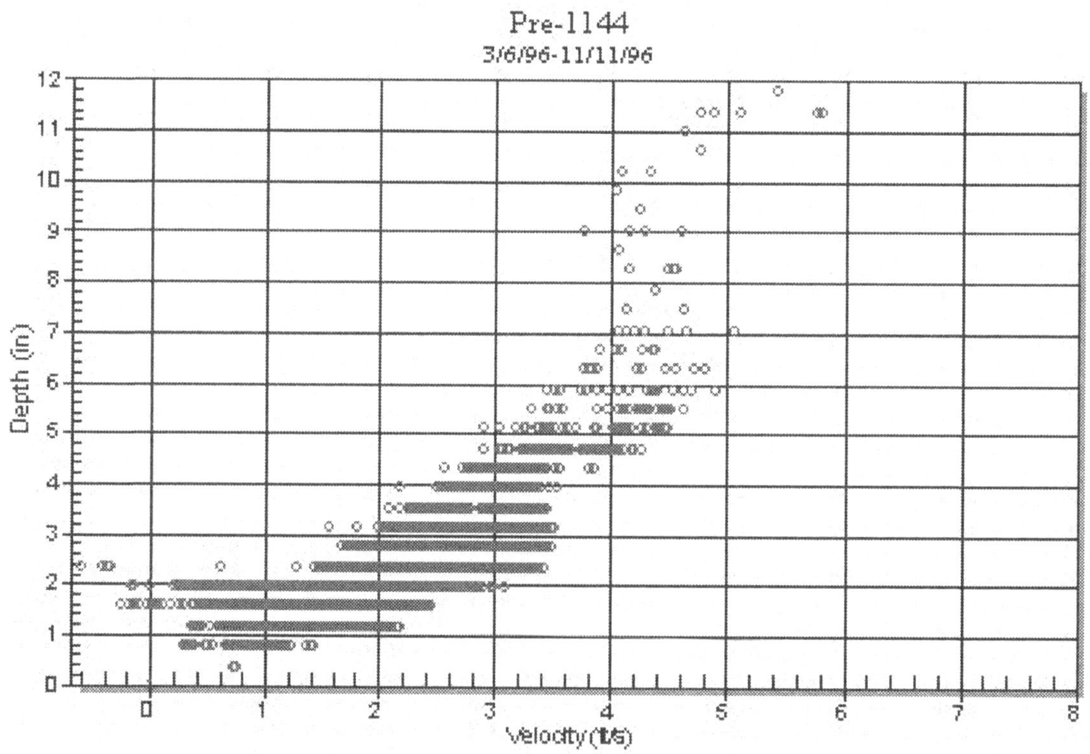

Figure F-3. Basin 1144: Velocity-Depth Scatter Plot Pre-Rehabilitation Flow Meter.

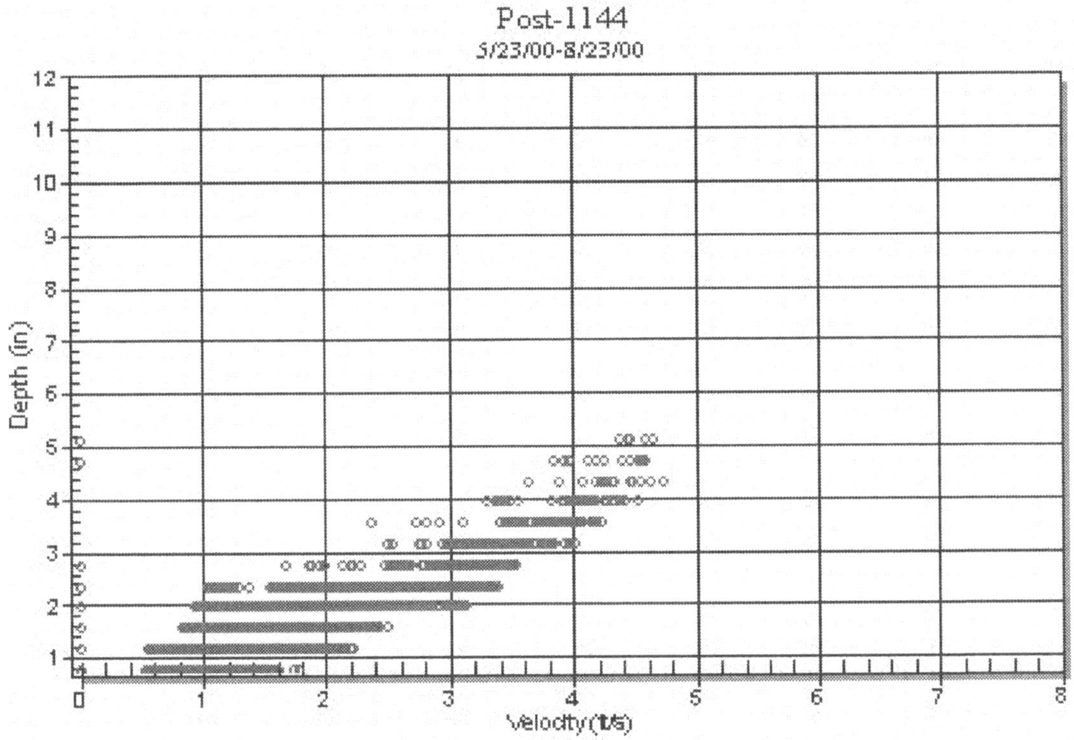

Figure F-4. Basin 1144: Velocity-Depth Scatter Plot Post-Rehabilitation Flow Meter.

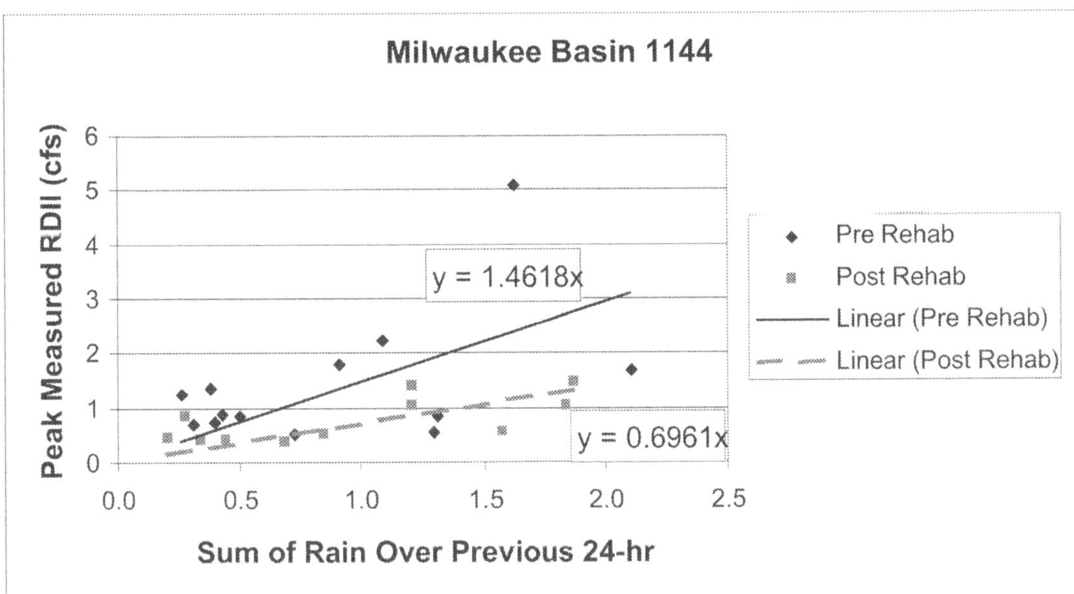

Figure F-5. Basin 1144: Plot of Peak Measured RDII vs. Sum of Rainfall in Previous 24 Hours.

REFERENCES

Agular, L. 1998. Contractor solves mystery of surcharged sewer system. *Water Eng. Management* 145(10):34–35.

Arbour, R. and K. Kerri. 1998. Collection systems: methods for evaluating and improving performance. California State University, Sacramento Foundation, Sacramento, California.

Bates, E.S. and R.H. Wynne Jr. 1989. Murphreesboro I/I reduction program works. *Tennessee Public Works Magazine.*

Bennett, D., R. Rowe, M. Strum, D. Wood, N. Schultz, K. Roach, M. Spence, and V. Adderley. 1999. *Sanitary sewer overflow (SSO) flow prediction technologies*, WERF Project 97-CTS-8.

Bergman, W. and R. Hardeman. 1991. Update on sanitary sewer rehabilitation in metropolitan Chicago. *Water Resources Planning and Management and Urban Water Resources,* J.L. Anderson (ed.). ASCE: New York, New York, pp. 825–829.

Bible, D. 1991. Results from sewer rehabilitation test areas. In: *Proceedings of the 64th WPCF Annual Conference*. Water Pollution Control Federation, Toronto, Ontario, Canada.

Brock, D.A. 1996. Storm period infiltration and inflow control. *Water. Env. Technol.* 8(11):51–53.

Brown, J.D. and M.D. Loehlein. 2001. Hunting ridge area wye grouting project.

Brown and Caldwell, Inc. 1995. Kent and Issaquah I/I pilot project: cost and effectiveness analysis. Unpublished results of study by King County, Washington, Dept. of Metropolitan Services, and Brown and Caldwell, Inc. engineers.

Brunner, P.A., R.B. Fernandez, J.S. Gresh, G. Knott, B.J. Schrock, H. Swetlik, and J. Warburton. 1994. Existing sewer evaluation and rehabilitation. *ASCE Manual and Reports on Engineering Practice No. 62.*

Cadden, R.G. and G. Mason. 1999. Project care: a holistic approach to wastewater system management. Presented at the *New Zealand Water and Waste Association Annual Conference*. New Zealand Water and Waste Association, Christchurch, New Zealand.

Cadden, R.G. 1997. Planning for the management of wet weather sewer overflow from separate sewer systems. Presented at the *Pipes Wagga Wagga Conference*. City of Wagga Wagga, New South Wales, Australia.

Cadden, R.G. 1999. Quality management of rehabilitation projects. Presented at the *Australia and New Zealand Rehab & Technologies '99 Conference*. Sydney, New South Wales, Australia.

Cadden, R.G. 2000. Where is there a place for trenchless technology in major wastewater system upgrades? Presented at the *Australasian Society for Trenchless Technology (ASTT)/International Society for Trenchless Technology (ISTT) 18th International No-Dig 2000 Conference*. ASTT/ISTT, Perth, Western Australia, Australia.

Capacity, Management, Operation and Maintenance Programs for Municipal Sanitary Sewer Systems. United States Environmental Protection Agency Office of Wastewater Management, 122.42(f). http://www.epa.gov/owmitnet/ccmom-5.htm.

CH2M Hill Engineering Ltd. 1990. Lot drainage characteristics study natural storm events. Prepared for the City of Edmonton, Alberta, Canada.

Conklin, G.F. 1981. Infiltration/inflow removal. *Water Eng. Management* 128(4):34–37.

Croxton, R. 1995. Manhole sealing stops severe I/I at Houston-area residential community. *Trenchless Technology* 4(3).

Davis, M. and C. Cleveland. 1999. Lott Basin OL22 removal effectiveness evaluation: task 1—determine I/I removal effectiveness final results. Brown and Caldwell, Inc. memorandum.

DeMonsabert, S. and P. Thornton. 1997. A benders decomposition model for sewer rehabilitation planning for infiltration and inflow planning. *Water Env. Res.* 69(2):162–169.

Dillard, W.C., G.E. Kurz, and P.A. Stonecipher. 1993. Management of sewer system rehabilitation for the overflow abatement program in Nashville, Tennessee. In: *Proceedings of the International Conference on Pipeline Infrastructure*. American Society of Civil Engineers, San Antonio, Texas, pp. 448–467.

Dorn, J. and S.W. Miles. 1997. Finding least-cost alternatives for controlling sanitary sewer overflows through application of STORM and a genetic algorithm. In: *Proceedings of the Annual Water Resources Planning and Management Conference*. American Society of Civil Engineers, Houston, Texas, pp. 750–755.

Driver, F.T. 1983. Manhole inflow and infiltration stopped with special repairs. *Water Eng. Management* 130(4):31–32.

Duggan, E., W. Goldberg, H. Woodfall, I. McSweeney, and D.J. Doherty. 1998. Planning and implementation of trenchless technologies to restore the St. James Avenue, Boston, interceptor. *Civil Engineering Practice* 13(2):77–86.

Earle, G.A., III, and R.P. Farrell Jr. 1997. Mathematical model for estimating sewer costs. *J. New England Water Env. Assoc.* 31:60–67.

Eng, T.H., G.A. Piggott, B.Y. Siong, A.W. Sharpe, and A.J. Killeen. 1994. Infiltration/inflow studies of Singapore sewerage catchments. Presented at the *Aqua Tech Asia International Conference on Asian Water Technology*. Aqua Tech Asia, Singapore, pp. 223–235.

Existing Sewer Evaluation and Rehabilitation—MOP FD-6. 1994. WEF.

Fernandez, R.B. 1986. Sewer rehab using a new subarea method. *Eng. and Management* 133(2):28–30.

Field, R. and T.P. O'Connor. 1997. Control strategy for storm-generated sanitary-sewer overflows. *J. Env. Eng.* 123(1):41–46.

Forbes, J.H. 1993. Wastewater collection. *Water Env. Res.* 65(6):303–305.

Fundich, A.E. and L.W. Kiest. June 1999. A case study in infiltration reduction through trenchless technology. *No-Dig International.*

Golden, J.B. 1996. Sanitary sewer overflows: an overview of U.S. EPA's national policy development. *J. New England Water Env. Assoc.* 30(1):39–48.

Gray, W.R. and R.J. Nogaj. 1990. Sanitary sewer bypass reduction program. *Water Eng. Management* 137(5):36–40.

Gustafsson, L.G., S. Winberg, and A. Refsgaard. 1997. Towards a distributed physically based model description of the urban aquatic environment. *Water Sci. Technol.* 36(8–9):89–93.

Gustafsson, L.G., S. Winberg, and A. Refsgaard. 1996. Towards a distributed physically based model description of the urban aquatic environment. In: *Urban Storm Drainage.* F. Sieker and H.R. Verworn (eds.).

Hilderhoff, S.R. and J.G. Wendle. 2001. Sanitary sewer rehabilitation by mini-basin to eliminate SSOs in the Paxton Creek drainage basin. Water Environment Federation.

Hollenbeck, A.J. and M. Rieger. 2001. Comprehensive public and private sector rehabilitation program provides documented 10-year storm protection. Water Environment Federation.

Hydrology Subcommittee of the Intragency Advisory Committee on Water Data, Bulletin 17B. 1982. USGS.

Juza, H.K., M.M. Vilhauer, and V.C. Adderley. 1996. Modeling and monitoring inflow reduction programs. In: *Advances in Modeling the Management of Stormwater Impacts.* W. James (ed.). Computer-Human Interaction, Guelph, Ontario, Canada, pp. 83–96.

Kailua and Kaneohe-Ahuimanu Smoke Testing Report, Barrett Consulting Group.

Keefe, P. 1998. Test basins for I/I reduction and SSO elimination. In: *Proceedings of the WEF Specialty Conference*, Water Environment Federation. Cleveland, Ohio, pp. 385–393.

Kidder, A.D., Jr. and W.H. Sukenik. 1997. Houston's solutions for successful rehabilitation. In: *Proceedings of the Conference on Trenchless Pipeline Projects.* American Society of Civil Engineers, Boston, Massachusetts.

Kidder, A.D., Jr., W.H. Sukenik, and T.L. Chance. 1997. The Houston experience: recipe for a successful demonstration project. In: *Proceedings of the Conference on Trenchless Pipeline Projects*. American Society of Civil Engineers, Boston, Massachusetts.

Kiefer, P.W. 1988. Greater mentor final SSES report. *CT Consultants.*

Kurz, G.E. 1997. Predicting I/I reduction for planning sewer rehabilitation. In: *Proceedings of the Conference on Trenchless Pipeline Projects.* American Society of Civil Engineers, Boston, Massachusetts, pp. 103–110.

Kurz, G.E., P.A. Stonecipher, C.W. Dillard, G.A. Ballard, L.B. Scott, and C.Q. Toosi. 1996. Successful sewer rehabilitation. *No-Dig Engineering*.

Lambert, G.D. and A.J. Hollenbeck. 1987. New approach achieves inflow reduction in sanitary sewers. *Public Works* 118(9):119–121.

Larsen, J.P., J.N. Struve, F. Bloetscher, and D. McLaughlin. 1997. After rehabilitation, then what? *Water Environ. Technol.* 9(4):44–49.

Loucks, E.D., and Johnson, T.Z. 1991. Hydraulic analysis of the impacts of I/I on system capacity. In: *Environmental Engineering Proceedings of the 1991 Specialty Conference on Environmental Engineering.* ASCE, New York, New York, pp. 114–119.

Macaitis, W.A. and R. Kuhl. 1994. Local sewer rehabilitation—metro Chicago. *URBAN Drainage Rehabilitation Programs and Techniques.* American Society of Civil Engineers: New York, New York, pp. 111–112.

Marlar, M. 1996. Historic army post tackles I/I. *Public Works,* 127(13):34–35.

Merrill, M.S. and R. Butler. 1994. New dimensions in infiltration/inflow analysis. Presented at the *WEF Wet Weather Issues Conference*. Water Environment Federation, Louisville, Kentucky.

Merrill, M.S., M. Davis, A. Lukas, and B. Topolsi. 2000. Impact of rehabilitation of different components of the sewer system on I/I reduction. Presented at the *Water Envrionment Federation Collection Systems Conference.* Water Environment Federation, Rochester, New York.

Merrill, M.S., A. Lukas, R. Swarner, and S. Klusman. 1997. Infiltration and inflow, infiltration or inflow. Which is the problem? Presented at the *United States Environmental Protection Agecy SSO Conference*. U.S. EPA, Washington, D.C.

Miles, S.W. and Aldrich, J.A. 1997. Demonstrated I/I reduction using trenchless methods. Presented at the *Ohio Water Environment Association Specialty Conference.* Ohio Water Environment Association.

Miles, W., B. Gullet, and D. Zimmer. 1996. Managing old sewers in the New South. *Water Environ. Technol.* 8(4):41–43.

Milwaukee Metropolitan Sewerage District Rules and Regulations. MMSD, adopted August 18, 1982, repealed and re-created September 28, 1998.

Moore, G.T., R.J. Neath, W.J. Wilson, M.F. Weishaar, and J.R. Emery. 1997. Modeling a sewer to develop SSO control. *Water Environ. Technol.* 9(7):33–37.

Nelson, R.E., Hsiung, P.H., and Witt, A.A. February 1999. *Optimization of Collection System Maintenance Frequencies and System Performance.* American Society of Civil Engineers, EPA Cooperative Agreement #CX 824902-01-0.

Park, H. and I.K. Lee. 1998. Existing sewer evaluation results and rehabilitation strategies: the city of Seoul, Korea. *Environ. Tech.* 19(7):733–739.

Popielarczyk, D.J., and Lenart, D.M. 1993. Elusive I/I: A case study of attempts to mitigate extraneous water problems in the North Brookfield sanitary sewer system. *J. New England Water Env. Assoc.* 27(2):130–149.

Public Works. 1996. "Beantown II" goes modern for sewer system study. 127(13):26–27.

Rosendaul, G., R. Engle, and B. Waite. 1996. Town stops over half of their total I & I. *Water Eng. Management* 143(8):23–24.

Rowlett, T. and K. Kelgard. 1992. When sewer rehab doesn't stop basement flooding. In: *Water Resources Planning and Management: Saving a Threatened Resource—In Search of Solutions, Proceedings of the Water Resources Sessions at the Water Forum.* ASCE, New York, New York, pp. 648–653.

Samson, J., G. Ju, G. Riek, and W. Sukenik. 1998. Analysis of sanitary sewer flow data and thorough demonstration of I/I reduction. *WEFTEC Extended Abstract.*

Sanitary Sewer Rehabilitation Demonstration Project. 1997. *City of Houston, Greater Houston Wastewater Program, Draft Final Report.* Executive Summary, Camp Dresser & McKee.

Sanitary Sewer Flow Monitoring and System Evaluation Report, Volume 1. Prepared for the City of Milwaukee, Wisconsin, by Rust Environment and Infrastructure, January 1997.

Sewer System Investigation—1997. Prepared for the City of Milwaukee, Wisconsin, by Rust Environment and Infrastructure, January 1997.

Sewer System Investigation—1998. Prepared for the City of Milwaukee, Wisconsin, by Rust Environment and Infrastructure, January 1998.

Sieger, R.B. and J.E. King. 1995. The limits of capacity. *Water Env. Technol.* 7(4):36–40.

Soil Survey of Waukesha and Milwaukee Counties Wisconsin. USDA (January 1998).

Spooner, C.S., J. Promise, and P.H. Graham. 1978. A demonstration of area-wide water resources planning. National Technical Information Service, Springfield, Virginia.

Steiner, M., J.A. Smith, S.J. Burges, C.V. Alonso, and R.W. Darden. 1999. Effect of bias adjustment and rain gauge data quality control on radar rainfall estimation. *Water Resources Research* 35(8):2487–2503.

Stephl, M. and J. Maciariello. 1993. Collection system successes. *Water Env. Technol.* 5(8):58–62.

Stevens, P.L. 1994. Improve your infiltration and inflow studies. *Water Env. Technol.* 6(7):46.

Sullivan, R.H., R.S. Gemmell, L.A. Schafer, and W.D. Hurst. 1977. Economic analysis, root control, and backwater flow control as related to infiltration/inflow control. Report EPA—600/2-77-017a, National Technical Information Service, Springfield, Virginia.

Swarner, R. and S. Klusman. 1995. I & I reduction in a metropolitan system: with multiple agencies, contractual constraints, and lack of incentive: is it worth the effort? Presented at *WEFTEC '95, WEF's 68th Annual Conference and Exhibition*. Water Environment Federation, Miami Beach, Florida.

Swarner, R. and M. Thompson. 1994. Modeling inflow and infiltration in separated sewer systems. Presented at *WEFTEC '94, WEF's 67th Annual Conference and Exposition.* Water Environment Federation, Chicago, Illinois.

2001 Sanitary Sewer Flow Monitoring. Prepared for the City of Milwaukee, Wisconsin, by Earth Tech, 2001.

2010 Facilities Plan, Volumes 1, 2, and 3. Prepared for MMSD, May 1998.

Vitasovic, Z., R. Swarner, and E. Speer. 1990. Real time control systems for CSO reduction. *Water Env. Technol.* 2–3:58–65.

Vogel, M. 1994. Rehab saves flooded basements. *Trenchless Technology* 3(6).

Wada, Y. and H. Miura. 1997. Application of a hydrologic model for excessive inflow/infiltration into a sewer pipe. *Technology Reports of Kansai University* 39:197–204.

Walch, M.P. and K.S. Leo. 1996. Miami-Dade water and sewer department's interactive weather radar and virtual watershed management. In: *Proceedings of Watershed 1996.* Water Environment Federation, Baltimore, Maryland.

Water Eng. Management. 1983. Experts discuss private sector infiltration/inflow. 130(10):32–42.

Wells, Jeff P. 1997. Service lateral rehabilitation technology overview. In: *Proceedings of the Conference on Trenchless Pipeline Projects*. American Society of Civil Engineers, Boston, Massachusetts, pp. 143–149.

Wilmut, C.G. and S.D. Sanders. 1990. Insights into I/I investigations. *Water Eng. Management*, 137(5):32–35.

Worldwater Env. Eng. 1996. Western New York community installs permanent flowmeter network for sewerage system management. 19(4):36.

Zheng, S. and V. Novotny. 1991. Stochastic modeling of combined sewer flows. *Water Sci. Technol.* 24(6):35–40.

Zimmerman, R.A. and R.D. Martin. 1993. From prevention to prediction. *Water Env. Technol.* 5(8):63–65.7

WERF SUBSCRIBERS

WASTEWATER UTILITY

Alabama
Montgomery Water Works & Sanitary Sewer Board

Alaska
Anchorage Water & Wastewater Utility

Arizona
Gila Resources
Glendale, City of, Utilities Department
Mesa, City of
Phoenix Water Services Department
Pima County Wastewater Management

Arkansas
Little Rock Wastewater Utility

California
Calaveras County Water District
Central Contra Costa Sanitary District
Contra Costa Water District
Crestline Sanitation District
Delta Diablo Sanitation District
Dublin San Ramon Services District
East Bay Municipal Utility District
El Dorado Irrigation District
Escondido, City of
Fairfield-Suisun Sewer District
Irvine Ranch Water District
Las Virgenes Municipal Water District
Lodi, City of
Los Angeles, City of
Los Angeles County, Sanitation Districts of
Napa Sanitation District
Orange County Sanitation District
Palo Alto, City of
Riverside, City of
Sacramento Regional County Sanitation District
San Diego Metropolitan Wastewater Department, City of
San Francisco, City & County of
San Jose, City of
Santa Barbara, City of
Santa Rosa, City of
South Bayside System Authority
South Orange County Wastewater Authority
Union Sanitary District
West Valley Sanitation District

Colorado
Boulder, City of
Colorado Springs, City of
Littleton/Englewood Water Pollution Control Plant
Metro Wastewater Reclamation District, Denver

Connecticut
New Haven, City of, WPCA

District of Columbia
District of Columbia Water & Sewer Authority

Florida
Broward, County of
Fort Lauderdale, City of
Gainesville Regional Utilities
JEA
Kissimmee, City of, Department of Water Resources
Miami-Dade Water & Sewer Authority
Orange County Utilities Department
Orlando, City of
Reedy Creek Improvement District
Seminole County Environmental Services
St. Petersburg, City of
Stuart Public Utilities
Tallahassee, City of
Tampa, City of
West Palm Beach, City of

Georgia
Atlanta Department of Watershed Management
Augusta, City of
Clayton County Water Authority
Cobb County Water System
Columbus Water Works
Fulton County
Gwinnett County Department of Public Utilities
Macon Water Authority
Savannah, City of

Hawaii
Honolulu, City and County of

Idaho
Pocatello, City of

Illinois
American Bottoms Wastewater Treatment Plant
Dupage, County of
Greater Peoria Sanitary District
Kankakee River Metropolitan Agency
Metropolitan Water Reclamation District of Greater Chicago
North Shore Sanitary District
Wheaton Sanitary District

Indiana
Fort Wayne, City of

Iowa
Ames, City of
Des Moines Metro Wastewater Reclamation Authority

Kansas
Johnson County Unified Wastewater Districts
Unified Government of Wyandotte County/Kansas City, City of

Kentucky
Louisville & Jefferson County Metropolitan Sewer District

Louisiana
Sewerage & Water Board of New Orleans

Maine
Bangor, City of
Portland Water District

Maryland
Anne Arundel County Bureau of Utility Operations
Howard County Department of Public Works
Washington Suburban Sanitary Commission

Massachusetts
Boston Water & Sewer Commission
Upper Blackstone Water Pollution Abatement District

Michigan
Ann Arbor, City of
Detroit, City of
Grand Rapids, City of
Holland Board of Public Works
Lansing, City of
Owosso Mid-Shiawassee County WWTP
Saginaw, City of
Wayne County Department of Environment
Wyoming, City of

Minnesota
Rochester, City of

Missouri
Independence, City of
Kansas City Missouri Water Services Department
Little Blue Valley Sewer District
Metropolitan St. Louis Sewer District

Nebraska
Lincoln Wastewater System

Nevada
Henderson, City of

New Jersey
Passaic Valley Sewerage Commissioners

New York
New York City Department of Environmental Protection

North Carolina
Cary, Town of
Charlotte/Mecklenburg Utilities
Durham, City of
Metropolitan Sewerage District of Buncombe County
Orange Water & Sewer Authority

Ohio
Akron, City of
Butler County Department of Environmental Services
Columbus, City of
Metropolitan Sewer District of Greater Cincinnati
Northeast Ohio Regional Sewer District
Summit, County of

Oklahoma
Tulsa, City of

Oregon
Clean Water Services
Eugene/Springfield Water Pollution Control
Water Environment Services

Pennsylvania
Philadelphia, City of
University Area Joint Authority, State College

South Carolina
Charleston Commissioners of Public Works
Mount Pleasant Waterworks & Sewer Commission
Spartanburg Sanitary Sewer District

WERF SUBSCRIBERS

Tennessee
Cleveland, City of
Knoxville Utilities Board
Murfreesboro Water & Sewer Department
Nashville Metro Water Services

Texas
Amarillo, City of
Austin, City of
Dallas Water Utilities
Denton, City of
El Paso Water Utilities
Fort Worth, City of
Gulf Coast Waste Disposal Authority
Houston, City of
San Antonio Water System
Trinity River Authority

Utah
Salt Lake City Corporation

Virginia
Alexandria Sanitation Authority
Arlington, County of
Fairfax County Virginia
Hampton Roads Sanitation District
Henrico, County of
Hopewell Regional Wastewater Treatment Facility
Loudoun County Sanitation Authority
Lynchburg Regional WWTP
Prince William County Service Authority
Richmond, City of

Washington
Edmonds, City of
Everett, City of
King County Department of Natural Resources
Seattle Public Utilities
Yakima, City of

Wisconsin
Green Bay Metro Sewerage District
Madison Metropolitan Sewerage District
Milwaukee Metropolitan Sewerage District
Racine, City of
Sheboygan Regional Wastewater Treatment
Wausau Water Works

Australia
South Australian Water Corp.
Sydney Water Corp.
Water Corp. of Western Australia

Canada
Greater Vancouver Regional District
Toronto, City of, Ontario

Mexico
Servicios de Agua y Drenaje de Monterrey, I.P.D.

United Kingdom
Yorkshire Water Services Limited

STORMWATER UTLITY

California
Los Angeles, City of, Department of Public Works
Monterey, City of
Santa Rosa, City of

Colorado
Boulder, City of

Georgia
Griffin, City of

Kansas
Overland Park, City of

Kentucky
Louisville & Jefferson County Metropolitan Sewer District

Maine
Portland Water District

North Carolina
Cary, Town of
Charlotte, City of, Stormwater Services

Pennsylvania
Philadelphia, City of

STATE

Fresno Metropolitan Flood Control District, Calif.
Urban Drainage & Flood Control District, Colo.

CORPORATE

ADS Environmental Services
The ADVENT Group Inc.
Alan Plummer & Associates
Alden Research Laboratory
Alpine Technology Inc.
Aquateam–Norwegian Water Technology Centre A/S
BaySaver Inc.
BioVir Inc.
Black & Veatch
Boyle Engineering Corporation
BPR CSO
Brown & Caldwell
Burns & McDonnell
CABE Associates Inc.
The Cadmus Group
Camp Dresser & McKee Inc.
Carollo Engineers Inc.
Carpenter Environmental Associates Inc.
CDS Technologies Inc.
Chemtrac Systems Inc.
CH2M HILL
Clancy Environmental Consultants Inc.
Damon S. Williams Associates, LLC
David L. Sheridan, P.C.
Earth Tech Inc.
Ecolab Water Care Services
EMA Inc.
The Eshelman Company Inc.
Finkbeiner, Pettis, & Strout Inc.
Frontier Geosciences Inc.
ftn Associates Inc.
Gannett Fleming Inc.
GE Betz
Golder Associates Inc.
Greeley and Hansen LLC
The HACH Company
Hazen & Sawyer, P.C.
HDR Engineering Inc.
HNTB Corporation
HydroQual Inc.
Insituform Technologies Inc.
Institute for Environmental Technology & Industry, Korea
Jacobson Helgoth Consultants Inc.
Jason Consultants Inc.
Jordan, Jones, & Goulding Inc.
KCI Technologies Inc.
Kelly & Weaver, P.C.
Kennedy/Jenks Consultants
Komline Sanderson Engineering Corporation
Lawler, Matusky & Skelly Engineers, LLP
Limno-Tech Inc.
Lombardo Associates Inc.
Malcolm Pirnie Inc.
McKim & Creed
MEC Analytical Systems Inc.
MWH
New England Organics
Odor & Corrosion Technology Consultants Inc. (OCTC)
ONDEO Degremont Inc.
Oswald Green, LLC
PA Government Services Inc.
Parametrix Inc.
Parsons
Post, Buckley, Schuh & Jernigan
R&D Engineering/Conestoga Rover & Associates
The RETEC Group
R.M. Towill Corporation
Ross & Associates Ltd.
Royce Technologies
SAIC Maritime Technical Group
Sear-Brown
Sierra Environmental Services
SYNAGRO
Tetra Tech Inc.
Trojan Technologies Inc.
URS Corporation
USFilter NATC
Vortechnics Inc.
Wade-Trim Inc.
Weston Solutions Inc.
Woodard & Curran
WRc/D&B, LLC
WWETCO, LLC

INDUSTRY

American Electric Power
ChevronTexaco Energy Research & Technology Company
The Coca-Cola Company
Dow Chemical Company
DuPont Company
Eastman Kodak Company
Merck & Company Inc.
ONDEO Services
Procter & Gamble Company
PSEG Services Corp.
Severn Trent Services Inc.
Shell Global Solutions
Thames Water Plc
United Water Services LLC

Note: List as of 5/12/03

Board of Directors

Research Council

Stormwater Technical Advisory Committee

WERF Product Order Form

s a benefit of joining the Water Environment Research Foundation, subscribers are entitled to receive one complimentary copy of all final eports and other products.Additional copies are available at cost (usually $10).To order your complimentary copy of a report, please write ree" in the unit price column.WERF keeps track of all orders.If the charge differs from what is shown here, we will call to confirm the total efore processing.

ame _______________ Title _______________

rganization _______________

ddress _______________

ity _______________ State _______________ Zip Code _______________ Country _______________

hone _______________ Fax _______________ Email _______________

Stock #	Product	Quantity	Unit Price	Total

Postage & Handling	
VA Residents Add 4.5% Sales Tax	
Canadian Residents Add 7% GST	
TOTAL	

Method of Payment:

(All orders must be prepaid.)

❑ Check or Money Order Enclosed

❑ Visa ❑ Mastercard ❑ American Express

ccount No. _______________ Exp. Date _______________

ignature _______________

Shipping & Handling:

Amount of Order	United States	Canada & Mexico	All Others
Up to but not more than:	Add:	Add:	Add:
$20.00	$5.00*	$8.00	50% of amount
30.00	5.50	8.00	40% of amount
40.00	6.00	8.00	
50.00	6.50	14.00	
60.00	7.00	14.00	
80.00	8.00	14.00	
100.00	10.00	21.00	
150.00	12.50	28.00	
200.00	15.00	35.00	
More than $200.00	Add 20% of order	Add 20% of order	

*minimum amount for all orders

Note: Please make checks payable to the Water Environment Research Foundation.

To Order (Subscribers Only):

Log on to www.werf.org and click on the "Product Catalog."

Phone: (703) 684-2470
Fax: (703) 299-0742.

WERF
Attn: Subscriber Services
635 Slaters Lane
Alexandria, VA 22314-1177

To Order (Non-Subscribers):

Non-subscribers may be able to order WERF publications either through WEF (www.wef.org) or IWAP (www.iwapublishing.com).Visit WERF's website at www.werf.org for details.

IWA Publishing's authorised EU representative for General Product Safety Regulations is Diane D'Arras, 15 rue Duret, 75116 Paris, France, e-mail: safety@iwap.co.uk.

Printed and bound by CPI Group (UK) Ltd, Croydon, CR0 4YY
12/05/2026
02108879-0013